OXFORD MEDICAL PUBLICATIONS

Critical Reading for Primary Care

6. The consultation: an approach to learning and teaching
   David Pendleton, Theo Schofield, Peter Tate, and Peter Havelock
9. Modern obstetrics in general practice
   Edited by G. N. Marsh
11. Rheumatology for general practitioners
   H. L. F. Currey and Sally Hull
15. Psychological problems in general practice
   A. C. Markus, C. Murray Parkes, P. Tomson, and M. Johnston
17. Family problems
   Peter R. Williams
18. Health care for Asians
   Edited by Bryan R. McAvoy and Liam J. Donaldson
19. Continuing care: the management of chronic disease (second edition)
   Edited by John Hasler and Theo Schofield
20. Geriatric problems in general practice (second edition)
   G. K. Wilcock, J. A. M. Gray, and J. M. Longmore
21. Efficient care in general practice
   G. N. Marsh
22 Hospital referrals
   Edited by Martin Roland and Angela Coulter
23. Prevention in general practice (second edition)
   Edited by Godfrey Fowler, Muir Gray, and Peter Anderson
24. Women's problems in general practice (third edition)
   Edited by Ann McPherson
25. Medical audit in primary health care
   Edited by Martin Lawrence and Theo Schofield
26. Gastrointestinal problems in general practice
   Edited by Roger Jones
27. Domiciliary palliative care
   Derek Doyle
28. Critical reading for primary care
   Edited by Roger Jones and Ann-Louise Kinmonth
29. Research methods for general practice (second edition)
   David Armstrong and John Grace

Critical Reading
for
Primary Care

Oxford General Practice Series • 28

Edited by

ROGER JONES

Wolfson Professor of General Practice, United Medical and Dental Schools (Guys and St. Thomas's), London

and

ANN-LOUISE KINMONTH

Professor of Primary Medical Care, University of Southampton

OXFORD   NEW YORK   TOKYO
OXFORD UNIVERSITY PRESS
1995

Oxford University Press, Walton Street, Oxford OX2 6DP

Oxford   New York
Athens   Auckland   Bangkok   Bombay
Calcutta   Cape Town   Dar es Salaam   Delhi
Florence   Hong Kong   Istanbul   Karachi
Kuala Lumpur   Madras   Madrid   Melbourne
Mexico City   Nairobi   Paris   Singapore
Taipei   Tokyo   Toronto
and associated companies in
Berlin   Ibadan

Oxford is a trade mark of Oxford University Press

Published in the United States
by Oxford University Press Inc., New York

A catalogue record for this book is available from the British Library

Library of Congress Cataloging in Publication Data
Critical reading for primary care / edited by Roger Jones and Ann-Louise Kinmonth.
p.   cm. — (Oxford general practice series ; no. 28)
Includes bibliographical references and index.
1. Medical literature — Evaluation.   2. Physicians (General practice — Study and teaching
(Continuing education)   3. Medicine — Research — Evaluation.   4. Family
medicine.   I. Jones, Roger, Prof.   II. Kinmonth, A.-L. (Ann-Louise)   III. Series.
R118.6.C75   1995   610'.71'5 — dc20   94-28163
ISBN 0 19 262381 8

Typeset by Colset Pte Ltd, Singapore
Printed and bound in Great Britain by
Biddles Ltd, Guildford and King's Lynn

175195

AAY-9872

Foreword

David L. Sackett, FRSC, MD, MSC Epid, FRCPC
Professor of Clinical Epidemiology
Director, Centre for Evidence-Based Medicine, Nuffield
Department of Clinical Medicine, University of Oxford

Ann-Louise Kinmonth and Roger Jones designed this book to 'introduce the reader to the existing range of research methodologies and illustrate their application in general practice,' and 'to attract more health professionals ... and research scientists to put their minds to work in the rich field of primary care.' Only time will inform us of their ultimate success; for now, I am sufficiently optimistic about the impact of what they and their colleagues have created to welcome their invitation to write a foreword for it.

My optimism is simple: effective learning is about mastering processes, not amassing facts, and this book excels in process! Reading it is like shadowing a series of gifted clinicians who talk to themselves as they read around interesting and sometimes troublesome patients. For it lets us inside the heads of both the clinical and non-clinical authors as they generate critical conclusions about the validity and clinical applicability of a collection of articles on the cause, presentation, perception, management, and outcome of human diseases, illnesses, and predicaments. Just as we learn best from gifted authors when they expose or 'talk through' their formation, refutation, and ultimate acceptance of diagnostic hypotheses, this book exposes how a group of gifted authors conduct their 'critical appraisal' of published evidence for its closeness to truth and practical usefulness.

The result will reward thoughtful readers in three ways. Firstly, they will learn a set of useful 'readers' guides' for critically appraising published evidence. When added to three other skills — the ability to convert uncertainty around a patient into an answerable question, the ability to find the best evidence to answer that question, and the ability to convert its critical appraisal into clinical action — the reader will become equipped to practice evidence-based medicine. In this vital dimension, the book doesn't risk becoming 'dated;' its evidence may go back 20 years, but its methods for appraising that evidence remain current. Moreover, by maintaining its focus on the processes of critical appraisal, it avoids losing its message even when an occasional conclusion of that appraisal is in error.

The second reward for readers ought to be their realization that applied, patient-based research in general practice is possible to do, possible to do well, and capable of answering some very important questions that cannot be answered anywhere else.

The final reward for thoughtful readers is the demonstration that good science is compatible with good manners. Each of the authors of the critically-appraised papers received a draft of that appraisal and an invitation to respond to it. That only one citation was lost in the course of this exchange is a tribute to the authors on both sides.

Contents

Contributors

John Bain Professor of General Practice, University of Dundee

Angela Coulter Director, King's Fund Centre, London

George Freeman Professor of General Practice, Charing Cross and Westminster Medical School, London

Andrew Haines Professor of Primary Health Care, University College of London Medical School, London

David Jewell Senior Lecturer in General Practice, University of Bristol

Roger Jones Wolfson Professor of General Practice, United Medical and Dental Schools, (Guy's & St. Thomas's), London

Ann-Louise Kinmonth Professor of Primary Medical Care, University of Southampton

Paul Kinnersley Lecturer in General Practice, University of Wales College of Medicine, Cardiff

David Mant Professor of Primary Care Epidemiology, University of Southampton

Elizabeth Murphy Lecturer in Social Studies, University of Nottingham

Penny Owen General Practitioner, Llanedyrn Health Centre, Cardiff

Roisin Pill Senior Research Fellow in General Practice, University of Wales College of Medicine, Cardiff

Martin Roland Professor of General Practice, University of Manchester

Bonnie Sibbald Senior Research Fellow in General Practice, University of Manchester

Claire Wilkinson Senior Lecturer in General Practice, University of Wales College of Medicine, Cardiff

1 Critical reading for primary care

Ann-Louise Kinmonth and Roger Jones

INTRODUCTION

This collection of critical readings arose from many discussions between ourselves, the authors, and others in general practice. We wanted to bring together a series of papers that would introduce the reader to the existing range of research methodologies and illustrate their application in general practice, if possible by primary care professionals and researchers themselves. We aimed to use these selected papers to encourage the development of a systematic approach to the primary care literature in general.

We invited contributions from many colleagues involved in primary care research, asking them to select for critical comment a paper that they felt had made, or should have made, an impórtant contribution to general medical practice. The responsibility for the final selection was ours, but we are indebted to our collaborating authors for their suggestions and for the professional and often very personal way they have demonstrated their appreciation of each paper and shown that a critical approach to reading the literature is significantly different from a wholesale demolition exercise.

We encouraged our collaborators to explain why they selected a particular paper for their contribution·and to provide a critique not only of the paper selected, but of the research method in general, including pointers to other relevant literature in order to present the principles for evaluating other studies using the same methodology. In addition, the authors of the original articles were offered the opportunity to comment on the critical reading of their articles. This process was not without incident – while most authors were appreciative, others returned critical readings of the critical readings. However, in only one case was a chapter lost in this way.

During the selection process, it soon became clear that some methods have been employed more frequently by general practitioners in their research than others; cross-sectional surveys were common and are exemplified by the work of Wallace and Haines, and Howie and colleagues. It was harder to find primary care papers exemplifying qualitative research methodology such as that by Morgan and Watkins, and difficult to identify an appropriate case control study. We have none the less been able to assemble relevant examples of all the main research methodologies carried out in general practice, if not always by primary care researchers.

The early chapters deal with qualitative research methods aimed at understanding the range of beliefs and responses of patients to disease. We have begun with these methods because of their particular relevance to the perspective of our discipline and because for many readers they may offer the excitement of discovering new ways of looking at old problems, but through an approach whose rules may be relatively new. This perspective of primary care includes a concern with the experience of illness as much as with the disease process; with the meaning of symptoms to the patient rather than with the results of sophisticated investigations. Qualitative research methods uniquely inform this perspective, and the chapters describing an approach to their critical assessment will assist the reader to avoid common pitfalls such as demanding initial hypotheses or population-based sampling techniques in studies whose aim is to understand, for example the belief systems underlying a decision by patients to stop taking prescribed medication. In addition, qualitative research recognizes the importance of the beliefs and priorities of the researcher as well as the researched and makes this explicit, as in Bain's treatment of the work of Thomas.

The chapters by Bain and Coulter take up the theme of the importance of dealing with the patient's experience of illness both in defining a case and in developing appropriate outcome measures. Subsequent chapters deal with the range of methods used to study association, causation, and efficacy. They include cross-sectional survey methods, a case-control study, and a cohort study. Small and large scale randomized controlled trials are discussed and the principles of evaluation of a diagnostic test are described. Throughout these chapters the authors deal with the problem of defining the research question, selection of the most relevant methodology, definition of a case, selection of the sample, appropriate outcome measures and their reliability, the effects of response rates, identification of bias and confounding, and issues of quality assurance. They also give general guidance on the interpretation of data and its implications for personal practice. The individuality of approach illustrates how personal perspective and experience necessarily colour the interpretation and meaning of any study and that it is not just application of the rules of critical assessment but the ensuing judgement as to the weight to be given to a particular set of results that matters. Hence the importance that is attached to the context of each study, be it qualitative or quantitative, and the emphasis given by authors to the associated literature.

READING THE LITERATURE

General practitioners face a considerable challenge in using the literature to keep abreast of innovation and change, in continuing medical education, in personal and practice development, and nowadays in career progression and in participation in audit and research. At one extreme, papers of increasing complexity and sophistication are generated by specialists working in important, narrowly defined areas of medicine while, at the other extreme, general practitioners are inundated by review, abstract, and 'topical' material, much of which is published for explicitly commercial motives and is of uncertain educational and scientific value. The middle ground is occupied by an increasing body of research conducted in primary health care and published by a relatively small number of peer-reviewed journals. We may need to turn to any one of these sources at different times and for different reasons, and ways in which we can usefully approach the literature for different purposes are discussed in more detail in the following chapter.

AN INVITATION TO RESEARCH

The further aim of this book is explicitly to attract more health professionals, those in allied disciplines, and research scientists to put their minds to work in the rich field of primary care. The papers selected illustrate a wide range of themes which are a continuing challenge to primary care research and point the way to the rich array of important questions worth tackling. For example we believe that the contributions on the importance of researching the views of patients and health professionals to understand better their behaviour both inside and outside the consulting room foreshadow an explosion of research in this area, as social scientists and practising health professionals collaborate more closely.

Several of the research publications included in this book predated and could have informed our current terms and conditions of service. These include responsibilities for health promotion, optimal list sizes, and out of hours calls. The papers chosen also show an important concern of primary care not to do harm. Examples include unacceptable intrusion into patients' lives (Chapter 4) over-zealous investigation (Chapter 5) with untested tests (Chapter 10), or the use of treatments that can do more harm than good, such as prescribing oral contraceptives for older women who smoke heavily (Chapter 8), bendrofluazide for men with mild hypertension (Chapter 9), or transdermal glyceryl trinitrate for patients with angina (Chapter 11). Many of the data contributing to clinical judgements in general practice are still derived directly from observations in the hospital setting. Since the costs and benefits of both diagnostic and therapeutic activities vary with the prevalence and severity of the disease, as well as the behaviour of patients, decisions on the best management strategies for primary care must derive from study of the natural history of illness and disease in that setting.

The consultation remains at the heart of our work. Thomas' research reminds us of the importance of asking research questions at the level of uncertainty with which we work in our daily lives. Despite methodological difficulties, the need for further description of safe, non interventionist methods of managing the 'worried well' is clear, in an era of crowded surgeries, too little time for the ill patient, a large and ever growing drug bill, increasing iatrogenic disease, and the proliferation of illness. If Thomas pursues the theme that patients can get better without prescription, Howie adds to the literature maintaining that patients can do better if they are given more time, and that shortage of time constrains quality of care. Time costs money, and further research is needed to delineate the areas in which the time of different health professionals can be spent most cost effectively.

Perhaps the main lessons for novice researchers to take from this book are that their ideas and energy are needed and valuable and that much as those of us who are clinicians are learning to serve our patients better by playing our part within the primary care team, so as researchers we will serve future patients best by playing our part within a multidisciplinary research team.

The last fifteen years or so have seen an almost exponential increase in research activity in and on general practice, in parallel with a growing self-confidence in the distinctive nature of the discipline, with consolidation of its academic base and with an increasing acceptance of the importance of undergraduate and postgraduate teaching and training. Traditionally, general practice research has been a minority activity for a number of reasons. Career progression did not depend on research achievement, or even on postgraduate qualification, and the terms and conditions under which general practitioners worked (and still work) incorporate positive disincentives to undertaking research. In this book, Coulter in particular comments on the constraints on general practice research.

That the road of the self-taught general practitioner researcher is difficult is incontrovertible; protected time, funding, and support can be difficult to obtain and GP researchers frequently find themselves emotionally isolated, not only by their hospital colleagues but also by partners in practice, who have other priorities and may not place much value on research work. Moreover, there have not been ready opportunities to work with the social and behavioural scientists and statisticians, on whose efforts much of the basic science of our discipline depends.

In recent years there have been several developments likely to improve the opportunity for research in primary care. The National Health Services research and development strategy is now in existence with a strong emphasis on multidisciplinary research in primary care; the review of the MRC boards has led to a broader constitution of the Health Services Research Board. The University departments of primary care have been underpinned by an interim financial award as a broad Service Increment for Teaching and Research (SIFTR) equivalent which for the first time allows them sufficient staffing to offer some support for research training within primary care teams, as well as their teaching functions.

There are in addition, many opportunities to contribute to research at a multi-practice level. Haines points out the peculiar strengths of British general practice for collaborative epidemiological research. These include the almost universal registration of individual patients with a general practitioner, allowing good denominator ascertainment and the gatekeeper role allowing accurate measurement of access to secondary care. The MRC framework and the RCGP Unit, whose work is discussed in the book, and local research networks all provide opportunities to become involved.

There is no room for complacency. The place of research and teaching in the career structure of general practitioners, other health professionals, and social and behavioural scientists working inside and outside academic departments of general practice still requires further thought and resources. Opportunities for NHS general practitioners and other members of the primary health care team at various stages of their careers to obtain research training and experience whilst retaining their commitment to a long-term career in NHS practice still need developing.

In the meantime, we hope that this book will whet your appetite for reading original articles and provide you with an informed approach to thinking about their meaning and contribution to the health of the patients we all serve, whether we are active in health care or research.

2 Approaching the literature

David Jewell and George Freeman

General practitioners face at least two significant problems in dealing with the medical literature. They must cope with the deluge of material that comes unbidden to their doors every day, while simultaneously using the literature as a source of study and information. When the judgement of history comes to be written in a few years' time, only 10–15 per cent of what gets published will prove to be of lasting value (Goffman 1981). However medicine changes constantly and an effort to keep abreast of change is required if we are to discharge our duties to our patients competently. For instance the recent past has seen the emergence of a completely new disease in the form of AIDS; new drug treatments for myocardial infarction (thrombolysis), renal stones (lithotripsy), breast cancer (tamoxifen), benign prostatic hypertrophy (finasteride); new forms of operative treatment such as minimally invasive surgery for cholecystectomy and endometrial ablation; alternatives to surgery such as RU486 for termination of pregnancy; and new forms of investigation such as MRI and PET scanning. We all need a strategy for extracting the essentials from the mountains of material available. In this chapter we try to give some guidance in terms of the overall strategy, to act as an introduction to the more detailed help and specific advice to be found in later chapters.

The best starting point would be a description of what we are reading at present. Unfortunately, there is little information on what general practitioners read and in any case, as with other aspects of general practice, we would expect a wide range of behaviour. General practitioners in the United Kingdom are likely to receive numerous free weekly and monthly magazines; they may subscribe to the weekly *British Medical Journal*, and *The Lancet*, the monthly *British Journal of General Practice* and the quarterly *Family Practice*, as well as other, more specialized publications. Commercial readership surveys suggest that more general practitioners read the free papers than the subscription journals, but they test reading and not recall, still less any effect on clinical practice. Such unsolicited material amounts to no more than getting information processed by others, although it may

---

* In a *dictionary of scientific quotations*, by Alan L Mackay, the full entry is 'Reading maketh a full man; conference a ready man and writing an exact man. [Bacon maketh a fat man — graffitol]'

alert us to an article of interest in another journal. Reading the original papers not only adds another dimension to the process, but also allows us to judge how selective others are being when they quote the literature. The Board of Examiners of the Royal College of General Practitioners recently introduced a 'Critical reading paper' as part of the written membership examination. This was, at least in part, a conscious attempt to influence trainees' habits and encourage them to read more, and it is clear that the effect has been considerable (Wakeford and Southgate 1992). Whether such habits will be maintained over a longer period remains to be seen. It is likely to depend on positive reinforcement from the benefits of reading on the reader, from the stimulus of reaccreditation, or both.

Reading, of course, is only a start. Without some system for retaining and then accessing and using the information the time spent reading could become an idle luxury. 'Reading the articles isn't a problem; it's remembering what I've read that is difficult', is a much repeated complaint. Learning theory gives some clues how to encourage retention. It is known that when we learn we use existing explanatory models already held in our memories (Klatsky 1980). New information is much more likely to be retained if it can easily be fitted into such models; paradoxically we may learn more from reading an article in our own area of expertise than from one in unfamiliar territory. Therefore for all new information and for anything else that might need to be accessed later it is important to have a system for retaining information and retrieving it at a later date. The system will depend entirely on the needs and habits of individuals, ranging from a box file for interesting articles, through manual index cards, to computerized databases. The advantage of keeping some sort of personal reference library is that for many of us the act of maintaining a system encourages both reading and retaining the information. The disadvantage is that it takes time. Other ways of making the information alive will help to retain it, and this might include discussing it with colleagues, presenting it at a journal club, or trying to put the conclusions into practice.

Whether we succeed in applying what we have read is the acid test. Again there is little information that indicates which strategies are most likely to encourage implementation. There is a wealth of data from audit studies which states very clearly how we all regularly fall short of our own internalized standards, but that in itself implies an aspiration to the highest standards of care. There are well known examples of doctors continuing with clinical practices after research has pronounced them useless or damaging, for instance the continuing prescription of diethylstilbœstrol in pregnancy some time after it had been shown not to be effective in preventing miscarriage (Chalmers 1993). However, clinical practice does change rationally, for instance the increasing use of prophylactic drugs in chronic asthma and reduction in the use of antibiotics in gastroenteritis. We know enough from studying our patients that merely supplying rational information is not sufficient to change behaviour on its own. Lomas has recently reviewed the evidence on what is required to change physicians' behaviour, emphasizing the importance of providing messages supplied by influential bodies, packaged in 'user friendly' format, through a variety of communication channels, with the backing of respected local exemplars, and with the opportunity to explore the implications of the research in personal encounters with local colleagues or respected outside authorities (Lomas 1994). Clearly some of these approaches can be incorporated into routine audit activities.

How we approach the literature, and what we do to retain the information depends on our aims. To illustrate this we have described three different levels of reading, and suggested different approaches for each one. These are *browsing*, in which the reader is trying to select interesting, useful, or stimulating material from routinely available publications; *reading for information*, in which the literature is used as a database of material to answer specific questions posed by the reader, usually related to everyday clinical practice; and *reading for research*, in which the reader is seeking to get a comprehensive view of the existing state of knowledge, ignorance, and uncertainty in a defined area by searching the existing literature under headings related to the topic of specific interest.

BROWSING

This section is concerned with the everyday reading of material that happens to cross one's path. It amounts to the relaxed browsing through the wealth of printed matter available routinely: research articles, editorials, abstracts, reviews, and even journalistic pieces in the free sheets. If our own experience is anything to go by, this is where most time is spent, and is what we rely on to keep ourselves up to date. The problem with reading at this level is that it can easily remain a passive routine of uncritical reading with little retention. Reading needs to be turned into an active process without going to the time and trouble entailed by rigorous application of the rules for critical reading, which are described later. Northedge's Good Study Guide (1990) sets out various strategies designed to keep you engaged by forcing you to think about what you are reading. First skim the article and ask yourself some questions. Such questions might concern the nature or relevance of the conclusions, or might (as with other levels of reading) concern the appropriateness of the methods used. While reading you should review the questions, and then reflect on them and the whole article at the end. Finally, writing some notes at the end can help to focus your mind on the conclusions and make sure you have understood it, as well as giving you a permanent record of what has been read.

Medical articles have the advantage that the abstract already provides the summary, making it much easier to complete the first task of thinking up questions to be answered. As experienced professionals, doctors will have lots of models in their heads as a basis for questions whether the article confirms, refutes, or adds to their existing body of knowledge. This body of personal knowledge works both ways. It will make the reading more active, but at the same time it will influence what people choose to read. Mostly it will encourage readers to concentrate on those articles that confirm or extend their own beliefs. For this reason we think that everybody should challenge themselves by consciously choosing the occasional article where the authors' views run counter to their own.

How should these articles be read? The guidelines for critical reading are further discussed in the rest of this book, and are summarized at the end of the book. However it is too much to expect that all casual reading should be done with that degree of intellectual rigour. Unfortunately, as anyone who has ever tried the exercise on published articles knows, it is unwise to trust the peer review system to do it for us. This is not because the peer review system doesn't work. Rather it is subject to all kinds of human failings, depending on the judgement of a small number of expert readers, all with their own prejudices and perspectives. Stephen Lock examined the system by following articles that had been rejected by the British Medical Journal and concluded that the peer review system worked less as a barrier and more as a traffic policeman, directing articles to the most appropriate place (Lock 1985). Like everything else, the system has its strengths in helping to improve the material that is published, but the judgement of what should be published is not only a scientific one. Editors fulfil their function by publishing articles that are important, and sometimes new and interesting, even though the content may be flawed. The pragmatic line, which we imagine most readers actually take, is to steer a path between the bland gullibility of believing everything, and the strenuous intellectualism of applying every rule, retaining a healthy scepticism for the literature. In particular beware the recycled conclusions printed in the free sheets, whose messages so often leave out all the qualifying clauses. As Sackett and colleagues (1991) pointed out, we give up more than clinical judgement when we let an authority tell us how to manage our patients. We also give up the opportunity to look at the actual clinical evidence to see whether it is both valid and applicable to our practice. Remember also that most published work will present its findings in the best possible light: even inveterate browsers should at least be aware of the checklists for critical reading and apply them once in a while.

Guidelines for browsing

1. Read the title. Is this something you might be interested in, that might be different from your own ideas, that might change your practice?
2. Read the abstract. If you are still interested think of one or two

questions you should ask while reading, for instance to check the validity of the conclusions, or to check that it really is generalizable to your own practice.

3. Read it, remembering to ask the questions you had in mind.

4. Make a note of your own conclusions. Try to discuss them with colleagues.

READING FOR INFORMATION

Professional life as a general practitioner is full of questions. What is the appropriate duration of treatment for acne rosacea? Is this funny rash the side effect of Azathioprine listed in the British National Formulary (BNF)? What really are the benefits (if any) of cervical cerclage in a patient having recurrent midtrimester miscarriage? Has this patient really got a carcinoid syndrome or am I just indulging every doctor's dream of diagnosing something rare? In one American study, physicians generated two such questions for every three patients seen (Covell *et al.* 1985). As well as questions concerning individual patients' problems, there are the others associated with audit, and the care and management of the whole practice. What standard of care should we be trying to achieve for our asthmatic patients, and how is it best organized? Can we agree on appropriate prescribing for patients with depression, and if so which of the vast range of antidepressant drugs available should we be using?

Textbooks can give some answers. They will usually represent the opinion of acknowledged experts in those fields who can act as guides to the reader, interpreting and linking the facts in the light of their own experience. The disadvantages of textbooks is that of necessity they cannot stay up to date, and this is most marked in fast changing areas. The writer's opinion may bias the selection of literature on which the conclusions are based. They do not always focus precisely on the question that the reader is asking. Where audit is concerned it is essential to use standards that are attainable in normal clinical work and not some theoretical ideal embodied in a textbook. Standards for audit also need to be set by those involved if the audit is to be successful; picking them from textbooks, off-the-peg, is unlikely to encourage strong commitment from the participants (Jewell 1992). Textbooks are organized information and can be seen as a pre-electronic database. Modern information technology will supplement them by offering easy, immediate access to many varieties of electronic database, regularly and rapidly updated.

The pioneering blueprint is the *Cochrane collaboration in pregnancy and childbirth* (CCPC), previously published as the *Oxford database of peri-*

natal trials (Enkin *et al.* 1993). This is a comprehensive survey of all randomized controlled trials in the field of obstetrics and perinatal medicine, grouped together so that they can be accessed by problem or clinical intervention. Each problem is then discussed by means of commentaries and meta-analyses. The intention is to provide the nearest to true answers that the literature will currently support. With the setting up of the Cochrane collaboration this work will gradually be extended into other fields. The *Cochrane collaboration in primary health care* was set up in 1993 with the intention of having material for publication by 1996.

For general practitioners desktop computers will soon give access to such information directly, possibly in the form of expert systems to support decision making for individual patient problems. There is a widespread fear that these advances, together with the agreement of guidelines for a variety of conditions, will supplant the personal element in general practitioners' decision making. Such systems should give patients and doctors the confidence that their decisions are based on clear evidence of the benefits and risks of different courses of action, where it is available. What the evidence means to individual patients in terms of personal risks and benefits, and how it is acted on will remain a matter for judgements made within the consulting room.

Where drug treatment of any condition is concerned, the best answer can be broken down to three questions. Is some treatment better than no treatment? Is one treatment better than another? Is a lot of treatment better than a little? With the array of drugs available, there is always a judgement to be made about the balance between supposed efficacy, convenience to the patient, side effects and cost. Definitive answers are therefore not easily available, and the need is for access to reasonably objective comparative articles to inform judgement. Pharmacy information services may be very helpful in providing information packs to those keen to discuss drugs. General practitioners in the United Kingdom now have the opportunity to receive detailed feedback of their own prescribing (PACT in England and Wales, standing for Prescribing, Analysis, and Cost; SPA in Scotland, standing for Scottish Prescribing Analysis), although lack of linkage with the clinical context will always be a limitation of these figures. Audit standards are set out in some published reports, and may be the distilled wisdom of an extensive literature review, or a comprehensive consultation exercise between generalists and specialists (Grol 1993). As experience and confidence with audit increases it is likely that more and more practitioners will be happy to share their results. Medical Audit Advisory Groups (MAAGs) should then be in a position to lead the development of local standards based on local circumstances.

The extent to which users are required to exercise critical faculties in this area of work will vary according to the nature of the task and the source of information. At one end, the *Cochrane collaboration in pregnancy and childbirth* is produced according to the highest standards of scholarship, providing a criterion reference point. At the other, a database of various practice based standards held by a MAAG will give a local reference based on a consensus of opinions of participating doctors. A selection bias may arise with information packs such as those concerned with therapeutics where selection is the key to the conclusions. Without the time to check the whole process, users will only be able to compare the pack with their own internal models and pay particular attention to obvious disagreements. Howie (1974) recognized this dichotomy years ago when he wrote that 'consensus may indicate a provisional standard for teaching and finding a lack of consensus may indicate a priority for further research.'

Guidelines for reading for information

1. Read the title, the authors, and their addresses. The title will tell you if you are in the right subject area. You may know something of the trustworthiness of the authors, their institutions, and track record. Beware, however, of relying too much on this 'halo' effect which could lead you to be too accepting of papers coming from well-known institutions and not accepting enough of others from less reputable ones.

2. Read the abstract. Not, as Sackett points out, to discover if the study is correct, but whether, if the conclusions turn out to be correct, you will be affected by the results.

3. Read the introduction, but only to ask a few limited questions. Is it clear what the study was setting out to do? Given that, and your own knowledge of the literature, are there any glaring omissions from their review of the current state of knowledge?

4. Go on to methods. Were they consistent with the aims of the study? How have the patient populations been chosen, and look at choice of controls, if any. Does the setting of the study affect the overall conclusions? Was the choice of outcome measures appropriate?

5. In results, check the figures, especially looking for loss of data from dropouts, and whether these were fully explained or not.

6. The discussion should cover two general questions. First, do the results really mean what they appear to mean, or are there reasons for doubting them. Second, how do these conclusions compare with the rest of the literature? More generally, are the authors trying to build too large an edifice of knowledge on an insubstantial base of evidence?

7. Finally, do you believe the overall message, and if not why not?

READING FOR RESEARCH

The disciplines of reading for formal research studies, in preparing review articles, applying for grants, or writing dissertations or theses for higher degrees are essentially the same as at other levels: we are approaching the literature with specific questions and critical faculties prepared. However, in addition to intellectual rigour, it demands a more systematic approach to the search, protected time, and a sound system of information storage and retrieval. The process might begin with published papers already known to the reader, using reference lists to work back through the literature. However at some stage it will be essential to search the literature methodically. This task has been made much easier by having Index Medicus on CD-ROM or Silver Platter, so that searches that previously required direct lines to distant MEDLINE facilities are now available on desktop computers. There are other databases in other fields, such as psychology, and citation indexes which are based on the frequency with which papers are quoted by others. It is a simple matter to search by subject heading, speciality, language, journal, time frame, and author. Enlisting the help of a librarian who will be more familiar with the techniques than most readers may make the task much easier, although the software is friendly enough for ordinarily computer literate amateurs to use successfully. Searches can very easily be combined; for example it is a simple matter to identify all papers published from family practice (the US equivalent of general practice) on the subject of the management of diabetes in the *Lancet*, the *British Medical Journal*, or the *British Journal of General Practice*. Having worked out such a search combination the sequence of commands can be saved on disk as a single word. This can then be used as a command to repeat the search strategy for successive years. Where the original paper contains an abstract the abstract will appear on screen, and the whole output for the searches can be printed onto paper or downloaded onto disk. To UK users Index Medicus has always had the drawback that its structure is determined by North American medicine. However, this is quickly overcome when one becomes familiar with the subject headings and search strategies.

Using MEDLINE

DJ was asked to write an article on practice nurses, and wanted to have a reasonably comprehensive overview of the literature (Jewell and Turton 1994). The department had recently installed the Silver Platter version of MEDLINE. This describes briefly the search strategy I used.

1. Start by inserting a Silver Platter compact disc in the disc transport. I used 1992, the most recent whole year available.

2. Find the appropriate subject headings. This system uses Thesaurus to explore the subject headings. 'Nursing' produces more than 200 subheadings. Of these, I selected Nursing care, Primary nursing care, Nurse practitioners, Community health nursing, Nursing services, Office nursing, and Home nursing.

3. The most likely looking heading is Community health nursing. So I highlight this one and press 'Search'. The first, welcome surprise is that it gives the definition of the term: 'A medical speciality concerned with the provision of continuing comprehensive primary health care for the entire family.' I seem to be on the right path. Searching on this yields 320 items. The screen shows:
   #1 320 COMMUNITY-HEALTH-NURSING/all subheadings

4. Repeat the process for Family Practice, the search yields 882. The screen shows
   #2 882 FAMILY-PRACTICE/all subheadings

5. This is the exciting part. To find how many of these specifically are about community health nursing in family practice, simply ask it to combine the two. So press Enter to start a new search, then type #1 and #2. The screen shows:
   #3 7 #1 and #2.   In other words, we are down to 7 articles in the whole year. One key brings these onto the screen, and they can be printed off if required. Most of these are obviously relevant and some look important. One of them is an article I had previously read, and should have been identified by the search, and two are in Danish.

6. Repeat the double search using Primary nursing care, Nurse practitioners, Nursing services, Office nursing, and Home nursing. The definitions are helpful in two. Primary nursing care is defined as 'Primary responsibility of one nurse for the planning, evaluation and care of a patient throughout the course of illness, convalescence and recovery.' In other words logical, but not what I expected. The search produces no articles for Primary Nursing care and Family Practice. Home nursing is just that, nursing in the home. The search identifies two articles in Family Practice but predictably they are irrelevant to practice nurses. Office nursing and Family Practice, Nursing services and Family Practice, and Nurse practitioners and Family Practice yield respectively 1, 1, and 9 articles. Most look important (one I was specifically hoping to find), and only one had already been identified in the first search.

7. So far the favoured search strategy could look like this:
   #1   COMMUNITY-HEALTH-NURSING/all subheadings
   #2   OFFICE-NURSING/all subheadings
   #3   NURSING-SERVICES/all subheadings
   #4   NURSE-PRACTITIONERS/all subheadings

#5   FAMILY-PRACTICE/all subheadings

#6   #1 or #2 or #3 or #4

#7   #6 and #5 — this yields 17 articles, all as previously identified. To check that this really does catch everything relevant, search on nursing care, which is the umbrella heading for many of the others.

#8   2193 NURSING-CARE/all subheadings. Working with large numbers like this is to be avoided if possible

#9   #8 and #5 gives 20 articles. This is worrying, since the three not identified by the previous strategy might be important. To examine just these three, use:

#10   #9 not #7. Relief that none of these three are of interest.

Finally, see if we can exclude the articles in foreign languages.

#11   Search on la(language) = English

#12   #7 and #11, brings it down to a manageable 15.

8. Here's the second exciting part. Now change the disc to the one for 1991. Then type #12. The computer repeats all the previous steps and comes up with 21 articles. I repeat the search back to 1988, when the system surprisingly, but correctly, comes up with one of my own articles. (Note: if you are using CD-ROM rather than Silver Platter you have to store the search strategy with a name on disk before changing the compact discs.)

This might look complicated. However using a system with which I was not entirely familiar the whole process took little more than an hour.

There are two omissions in Index Medicus. First there is research currently in progress. It may be possible to find out what work is being done by consulting registers of research. The Royal College of General Practitioners produces Research Intelligence, which lists all studies notified to the compilers. The entries do not show much detail, but it is possible to identify other people working in the field who may be happy to share ideas. Other ways of obtaining this information include contacting colleagues in university departments, who may be aware of ongoing research and general practitioners involved in local and regional research networks who similarly may be able to identify current research in your area of interest. Second there is the research which was completed but remained unpublished. This is more likely to worry those writing systematic reviews than those planning research projects. Publication bias describes the greater likelihood of publication of articles showing statistically significant results when compared with those showing no significant results (Easterbrook *et al.* 1991). Reviewing only published material may therefore overestimate the effects of a particular intervention. Writing systematic reviews is now becoming a much more rigorous process, and they might now be expected to include a clear statement of the question being addressed, a description of the search strategy followed to identify the primary research included, an assessment of the validity of the research identified, a discussion of the variation in the findings from different studies, and a description of the methods used to combine the different results (Oxman and Guyatt 1988).

For this level of reading the studies need to be read with considerable care. Apart from identifying flaws that need to be corrected in future work, it is important to ask questions such as how populations differ, about the shortcomings of the methods of measurement, the questions that should have been asked but were not, which results look misleading, either being chance findings (alpha errors) or artefacts. You will be interested in the balance of evidence, the boundaries of knowledge and ignorance, the size of reported effects, and the level of uncertainty. In general practice research particular attention needs to be paid to bias, particularly that introduced by selection of patients into studies, and that potentially introduced when they are large numbers of dropouts between the beginning and the end of a study. A general strategy is offered below, and more detail is given in the remainder of this book.

Such reading cannot be completed at one pass of the text. Like any in depth learning process an initial familiarization with the material will generate a provisional classification. This will inform the next reading which will in turn enable readers to get a more exact classification. Details may still need to be checked at subsequent readings of key passages. There are analogies here with qualitative research methods (Lincoln and Guba 1985).

Readers at this level will find it necessary to have a system of recording essential information to make meaningful comparisons between different sources. In essence this will amount to a refinement of the system described above, so that the questions are organized into a consistent format. Readers embarking on this sort of exercise for the first time are strongly advised to use an electronic format for their personal reference library. These can be constructed out of standard database packages, or by using specifically designed commercial packages such as Papyrus or Reference Manager. Such systems will save much repetition and simplify the process of linking studies. The available systems for both searching the literature and storing personal bibliographies has been reviewed by Jones (1993).

The most obvious reason for searching the literature at the beginning of a research study is to avoid unnecessary repetition of previous work. However, we must sound a note of caution; in many subject areas the literature is so extensive that one might never progress from the search to the research. Howie (1979) has counselled against too much reading early in the development of a research idea, rather seeing reading as a means of checking that development is along the right lines. The opinion of an expert can be an invaluable reassurance.

Surprising as it may seem, successfully refining an important but as yet unanswered research question, or synthesizing a number of single research papers into a coherent review, can be an invigorating and rewarding experience, particularly if you arrange protected time for the purpose.

Guidelines for critical reading (reproduced from Sackett *et al.* (1991) with permission) are given at the end of the book.

CONCLUSIONS

Having plumbed the depths of the most demanding reading, it is worth coming back to the surface to get a more comprehensive overview. We have suggested that good reading, like general practice itself, requires a wide repertoire of skills. Overall effectiveness will depend as much on the ability to select the appropriate skill for the appropriate task as on the exercise of the skills themselves.

We value and use each of the three reading levels described in this chapter. Browsing provides the best chance of meeting the unexpected and thus stimulating a new idea or connection. Information gathering should be much more effective when informed by the guidelines, while reading for research needs to be reserved for the tasks where the reader bears personal responsibility for the consequences of getting the best current spread of evidence about a chosen topic. Above all the 'good reader' needs to maintain a balance between being receptive to real advances which make the effort of ploughing through acres of newsprint rewarding, and a healthy, slightly conservative, scepticism for the transient novelties that will eventually amount to no more than tomorrow's fish and chip wrapper.

REFERENCES

Chalmers, I. (1993). Diethylstilbœstrol (DES) in pregnancy. In *Pregnancy and childbirth module* (ed. M. W. Enkin, M. J. N. C. Keirse, M. J. Renfrew, and J. P. Neilson), Cochrane database of systematic reviews: Review No. 02891, 3 September 1992. Published through '*Cochrane updates on disk*', Oxford: Update Software, Spring 1993.

Covell, D. G., Uman, G. C., and Manning, P. R. (1985). Information needs in office practice: are they being met? *Annals of internal Medicine*, **103**, 596–9.

Easterbrook, P. J., Berlin, J. A., Gopalan, R., and Matthews, D. R. (1991). Publication bias in clinical research. *Lancet*, **337**, 867–72.

Enkin, M. W., Keirse, M. J. N. C., Renfrew, M. J., and Neilson, J. P. (1993).

Pregnancy and childbirth module, Cochrane database of systematic reviews. *Cochrane updates on disk*, Oxford: Update Software.

Goffman, W. (1981). Ecology of the biomedical literature and information retrieval. In *Coping with the biomedical literature* (ed. K. S. Warren). Praeger, New York.

Grol, R. (1993). Development of guidelines for general practice care. *British Journal of General Practice*, **43**, 146–51.

Howie, J. G. R. (1974). *Clinical decision making in general practice* PhD thesis. University of Aberdeen.

Howie, J. G. R. (1979). *Research in general practice*. Croom Helm.

Jewell, D. (1992). Setting standards: from passing fashion to essential clinical activity. *Quality in Health Care*, **1**, 217–18.

Jewell, D. and Turton. C. (1994). What's happening to practice nursing? *British Medical Journal*, 308, 735–6.

Jones, R. (1993). Personal computer software for handling references from CD-ROM and mainframe sources for scientific and medical reports. *British Medical Journal*, **302**, 180–4.

Klatsky, R. (1980). *Human memory: structures and processes*. 2nd edn. W. H. Freeman and Co, San Francisco.

Lincoln, Y. S. L. and Guba, E. G. (1985). *Naturalistic enquiry*. Sage Publications, Beverley Hills, California.

Lock, S. (1985). *A difficult balance. Editorial peer review in medicine*. Nuffield Provincial Hospitals Trust, London.

Lomas, J. (1994). *Teaching old (and not so old) docs new tricks: effective ways to implement research strategies*. Sage Publications, Newbury Park, Canada (in press).

Northedge, A. (1990). *The Good Study Guide*. The Open University, Milton Keynes.

Oxman, A. D. and Guyatt, G. H. (1988). Guidelines for reading literature reviews. *Canadian Medical Association Journal*, **138**, 697–703.

Sackett, D. L., Haynes, R. B., Guyatt, G. H. and Tugwell, P. (1991). *Clinical epidemiology. A basic science for clinical medicine* (2nd edn). Little, Brown and Company, Boston.

Wakeford, R. and Southgate, L. (1992). Postgraduate medical education: modifying trainees' study approaches by changing the examination. *Teaching and learning in medicine*, **4**, 210–13.

3 The potential of qualitative research in primary care

Elizabeth Murphy

INTRODUCTION

The paper by Morgan and Watkins (1988) forms the focus for discussion in this chapter. It is one example of the growing number of studies, relevant to the concerns of general practice, which adopts qualitative interviews as its central methodology. Qualitative methods have played a central role in both medical anthropology and medical sociology, but it is only in relatively recent times that the discipline of general practice has begun to recognize and exploit their potential (McWhinney 1991).

For those trained in the school of quantitative method, qualitative research can be seen as 'soft' and 'lacking in rigour'. It may appear to ride roughshod over legitimate preoccupations with objectivity, control of extraneous variables, standardization, representativeness, and the like (Pope and Mays 1993). On the other hand, it has been argued that the methodological commitments typical of qualitative methods reflect the philosophical orientation of general practice itself (Murphy and Mattson 1992).

Quantitative and qualitative methods are best seen as complementary rather than competing paradigms for research in general practice (McWhinney 1991), as in other disciplines (Hammersley 1992). As Hammersley (1993) argues,

In doing research one is not faced with a choice between two well-defined routes that go off in opposite directions. Instead, the research process is more like finding one's way through a maze. And it is a rather badly kept and complex maze, where paths are not always distinct, where they wind back on one another, and where one can never be entirely certain that one reached the centre. (p. 56)

The context of the investigation, reported by Morgan and Watkins (1988), lies in research which demonstrates that a substantial proportion of those diagnosed as having hypertension either drop out of treatment, or take their prescribed medication sporadically. The authors attempt to throw light on this medically defined problem by eliciting hypertensive patients' own beliefs, concerns, and patterns of behaviour. They contrast this with previous attempts to understand patient 'non-compliance' which have focused upon the distribution of beliefs which are hypothesized to influence patients' responses to drug therapy. Also, whereas the prime concern of such studies is with patient 'default', the approach adopted by Morgan and Watkins reflects earlier sociological work (Stimson 1974), which suggests that the definition of patients as 'non-compliant' may stop us seeing patient behaviour as the outcome of rational decision-making based on people's own beliefs, evaluations, and circumstances.

STRENGTHS OF THE QUALITATIVE APPROACH

In contrasting their own approach with that of previous studies, Morgan and Watkins demonstrate the respective strengths of quantitative and qualitative approaches to this field of enquiry. Previous researchers have, with varying degrees of success, created causal models, leading to predictions which can be tested through empirical research. Morgan and Watkins, on the other hand, were attempting to describe the beliefs, concerns, and behaviour of hypertensive patients in the patients' own terms. They sought to explain responses to medical advice in terms of patients' own rationalities, rather than in terms of some externally created explanatory variables, devised by the researchers. This methodological orientation has, as the authors acknowledge, both strengths and weaknesses. It allows them to describe the beliefs and concerns which patients hold about hypertension and its management. It also gives access to the way in which the patients see these beliefs and concerns relating to their actions. Thus, the researchers do not have to infer causation from statistical relationships observed between sets of variables expected to discriminate between compliers and non-compliers, and the behaviour of patients. This approach allows us to substitute an understanding of patients' actions in their own terms, for the ability to predict the likely behaviour of certain groups of patients, based on our knowledge of their measurable beliefs, psychological profile, or material circumstances.

It is unhelpful to think of either the qualitative or the quantitative approach as fundamentally 'better' or 'worse' than the other. The appropriate question is surely which method yields more useful information for the particular task in hand. On the one hand building a causal model of compliance, based upon the findings of quantitative research, may be useful in informing health services planning. For example discovering that non-compliance is associated with a particular constellation of socio-demographic variables may assist planners to focus educational programmes upon groups where non-compliance is expected to be high. On the other, an understanding of the range of ways in which patients make sense of their diagnosis and formulate what is, to them, an appropriate response to it, is more likely to be useful to general practitioners whose aim is to negotiate the management of hypertension with individual patients. Research such as this can sensitize medical practitioners to the variety of ways in which lay people may make sense of medical conditions and help them to avoid the assumption that patients will necessarily share their doctor's models of the causes, nature, and appropriate management of medical conditions (Helman 1985). The need for such sensitizing is demonstrated clearly in this paper. In this study, patients made it clear to the researchers that they did not always make their concerns and reservations about the treatment they had been given explicit to their doctors.

This paper further demonstrates that, even where patients share the same linguistic terms as their doctors, one cannot assume that they share a common understanding of these terms. By exploring the various meanings attached to the term 'hypertension', Morgan and Watkins illustrate the potential for misunderstanding in lay-medical dialogue.

The researchers do not claim that their sample is in any way representative of those with hypertension, either locally or nationally. As in much qualitative, as well as quantitative, research in general practice, the sample appears to have been chosen for convenience and accessibility rather than any claim to representativeness. Traditionally qualitative researchers have paid scant regard to the generalizability of their findings. To some extent, this reflects the roots of qualitative research in anthropological studies where the concern was to describe the unusual and exotic aspects of traditional cultures rather than with a search for typicality. Added to this, as Schofield (1993) has pointed out, many aspects of qualitative research are inconsistent with achieving generalizability as traditionally conceptualized within the quantitative tradition. Schofield reports that a number of authors have recently attempted to reconceptualize generalizability in a way which is more relevant to qualitative studies. In particular, they have argued that the generalizability of qualitative research depends upon an assessment of the extent to which the situation studied matches other situations in which one is interested (Guba and Lincoln 1981, 1982).

To be able to assess this 'matching' it is vital that the research report should contain sufficient information about the particular cases studied to allow the reader to assess the 'comparability' of the research situation and that to which (s)he wishes to extrapolate the findings. One of the strengths of Morgan and Watkins' paper lies in such a detailed description of their sample. The careful description of both socio-demographic and clinical characteristics of respondents allows the reader to assess both the potential for, and the limits to, the generalizability of the analysis presented here. Since the authors are concerned to provide an analytic description of the beliefs and practices of a group of people with hypertension, which may offer insights which are relevant to others with this condition, this approach is appropriate. They are not here attempting to discover general probabilistic laws of human behaviour.

This paper demonstrates many of the strengths of research in the qualitative tradition. By encouraging respondents to talk freely, rather than attempting to impose a pre-determined structure on their talk, the researchers have maximized the possibility of discovering aspects of respondents' thinking or behaviour which they had not anticipated. Good examples of the fruitfulness of this approach are the discovery of the practice of 'leaving off' medication, particularly among the West Indian respondents, and the way in which respondents report testing out the effect of both orthodox medication and alternative remedies, against measurements of their blood pressure.

PROBLEMS WITH THE QUALITATIVE APPROACH

Alongside these strengths, this paper also raises some methodological problems. The method by which the sample was identified (through hypertension registers and doctor recall) had the disadvantage, acknowledged by the authors, of excluding those who have dropped out of treatment, completely. Since this group of patients represents a substantial proportion of those diagnosed as suffering from hypertension, the effect of this method is to exclude those whose behaviour is least consonant with the medical model of how those with hypertension 'ought' to respond to the diagnosis. If the authors had adopted a longitudinal, rather than a cross-sectional approach, they could have identified patients at diagnosis and then described the process by which some patients came to opt out of treatment while others accepted medication with varying degrees of commitment. Such a longitudinal study would, however, have been considerably more demanding in terms of time and resources, as well as being fraught with all the problems of sample attrition over time. General practice research is often the art of the possible and it may well be that the costs, both financial and practical, of a longitudinal study would have outweighed the benefits.

The researchers report that the research was based upon a 'schedule of open-ended questions and conducted in a relaxed, conversational manner'. There is no discussion of where the questions themselves came from. It is implicit, in the description of methods, that the questions were generated by the researchers in advance of the interviews and focused upon beliefs about the causes of hypertension, the meaning of the term hypertension, the symptoms of hypertension, the consequences of hypertension, and readiness to follow prescribed treatment for hypertension. In other words, the researchers set the agenda and defined what was relevant to their study in advance of the interviews. An even more fruitful approach might have involved encouraging respondents to describe their experiences around their diagnosis as 'a person with high blood pressure', before going on to explore these pre-defined areas in this systematic way. One of the justifications for the use of qualitative techniques is that the researcher often 'does not understand what (s)he does not understand' (Becker and Geer 1957), and starting interviews with a minimum of researcher-imposed structuring increases the possibility that unanticipated respondent 'ways of seeing' the situation will be identified.

For example Morgan and Watkins report that respondents were told that 'the general practitioner had given their name as a patient with high blood pressure'. The assumption appears to have been made that the patients themselves all shared this definition of themselves as 'people with high blood pressure', but no evidence is presented to corroborate that assumption. Qualitative interviews are above all an opportunity to examine a situation from the respondent's perspective, but in this case, by assuming that the patients shared the medical definition of their 'objective reality', Morgan and Watkins may have missed out on important dimensions of the respondents' subjective definitions of their situation.

The use of comparison has long been recognized as central to the logic of both quantitative and qualitative research (Hammersley 1993; Glaser and Strauss 1967). Morgan and Watkins have organized much of their analysis around a comparison between West Indian and European respondents. The majority of European respondents were English, but the sample also included four respondents from Southern Ireland and one from Spain. For the purposes of comparative analysis, all the European respondents were treated as a homogenous group and there is no discussion of the extent to which the behaviour and beliefs of the non-English Europeans does or does not conform to the pattern of the English respondents. Given the strong cultural identity of different European groups, it is perhaps surprising that Morgan and Watkins have chosen to combine these groups for the purposes of analysis (Payer 1990). In the light of the emphasis given to cross-cultural comparisons within the study, it might have been preferable to exclude all non-English respondents from the sample, in order to avoid the ambiguity which arises from the heterogeneity of the European sub-sample.

Morgan and Watkins report that almost half of their 60 respondents identified stress, worry, or tension as the cause of their high blood pressure. They discuss, at length, why this cause should be so popular in relation to hypertension, given that such a causal link has yet to be demonstrated in terms of the pathophysiological mechanisms of hypertension. They explore empirically the possibility that such causal theories may arise from lay misunderstanding of the term 'hypertension' and conclude that this is not an adequate explanation for their particular sample. They turn instead to the influence of general practitioners in reinforcing lay explanations of hypertension in terms of stress.

While lay and medical models do undoubtedly influence one another in the dynamic way which Morgan and Watkins suggest, it may be that, by focusing narrowly on hypertension, Morgan and Watkins have failed to observe the relationship between lay causal theories about high blood pressure and lay causal theories more generally. There may be no need to explain why lay people are so ready to see high blood pressure as being caused by stress. Indeed it would be more surprising if respondents had not attributed this condition to stress. The literature (Blaxter 1990; Calnan 1987; Murphy 1992) suggests that stress is a causal category which lay people are particularly ready to invoke, both in discussing illness at a general level and in relation to a wide range of specific disease diagnoses. This suggests that it is unnecessary to search for particular reasons why high blood pressure should be understood as the result of stress, but rather that it may be seen as reflecting a general readiness to invoke psychological explanations for illness.

One of the strengths of qualitative research methods lies in the scope which they offer for retaining the context of respondents' thinking. The practical dilemma often associated with this strength is that the context of any particular enquiry is, in principle at least, boundless. The difficulty with which researchers are forced to engage is that of deciding to what extent it is possible to retain the context of a particular enquiry. This is illustrated by our discussion of causal theories in relation to hypertension. Clearly there is a sense in which respondents' thinking about the causes of illness in general is a relevant context within which to study thinking about high blood pressure in particular. However, resources may simply not have been available to investigate respondents' broader thinking about health and illness alongside their thinking about high blood pressure in particular. In such cases the researchers are thrown back upon cumulative research and, in this particular case, our understanding of lay concepts of health and illness has grown substantially in the period since this study was carried out.

This discussion about the scope of any particular qualitative study is also relevant to the fraught question of sample size in qualitative research. Since the aim of such research is generally to provide a description of the range of respondents' understandings (rather than to count the number of informants falling into pre-specified categories and infer the characteristics of the population from the characteristics of the sample), the sampling criteria applicable to quantitative work are often not particularly helpful.

In qualitative studies the primary concern is with capturing the range of respondents' understandings. How then is one to decide what is an adequate sample size in qualitative research? Morgan and Watkins chose to interview 60 respondents. This study yielded data of considerable variety but it is doubtful whether the authors would claim that this represents an exhaustive study of lay understandings of hypertension and its management. No doubt if they had interviewed more respondents they might have identified additional aspects of respondents' thinking. One 'counsel of perfection' in qualitative research suggests that the decision about sample size cannot be made in advance (Glaser and Strauss 1967). Rather the researcher is committed to collecting data until a point of 'theoretical saturation' is reached. At this point no new insights or variations emerge from the study and the researcher may conclude that (s)he has exhaustively described the situation (s)he is interested in.

Such an approach is rarely feasible in practical terms. Few funding bodies are ready to sponsor such open-ended research and few researchers to commit themselves to it. Researchers are generally obliged to decide in advance how many respondents they wish to interview, given resource constraints and the importance of the area they are studying.

Given that it is estimated that one hour's interview represents approximately eight hours of a typist's time for transcription and a further twenty hours for detailed qualitative analysis, the decision to increase the sample size by even ten respondents has considerable implications for the time and cost involved in carrying out such research. In the context of limited resources, the choice in any given study often comes down to a trade-off between breadth and depth. One may have to choose between interviewing a large number of people in a relatively superficial way, or a much smaller number in greater depth. Since one of the strengths of qualitative interviewing lies in the scope for moving beyond superficial accounts and analyses, it is not surprising that qualitative researchers tend to opt for the latter option.

Morgan and Watkins do not give any details of how they carried out the analysis of the interview material. In spite of the recent development of computerized packages for the analysis of qualitative data, this process is still both complex and problematic and the constraints which operate in relation to the publication of findings do little to ease the situation. All too often, researchers find themselves developing rich analyses of their interview material, sensitive to the nuances of meaning, only to be forced to reduce these to a shadow of their former selves to fit publication criteria. Unlike those who report the findings of quantitative studies, qualitative researchers are rarely able to make use of summary statistics or tabulations to represent their data. Morgan and Watkins' paper is itself substantially longer than those generally published in medical journals, and is more typical of those published in social science journals. As the contribution of qualitative studies to general practice develops, journal editors will need to review their publication policies if such studies are to find their way into medical rather than social science journals.

In spite of the relative length of this paper, Morgan and Watkins have clearly had to be selective in the analyses they have chosen to present. Inevitably the reader is left with unanswered questions. I wanted to know how respondents actually conceptualized high blood pressure. We are told what they thought caused it, and what they believed the potential consequences were, but we do not discover what they actually believed it was. Likewise, we are told about the conventional and herbal remedies which respondents were taking for high blood pressure, as well as their lack of participation in yoga or relaxation classes. However, I wanted to know how respondents managed their condition at a less formal level. For example we are told that some respondents attributed their high blood pressure to diet and obesity, but we do not discover whether respondents were trying to manage their condition by changing their diets or reducing their weight.

To point to unanswered questions such as these should not be taken as a criticism of the paper. Practical constraints inevitably mean that researchers can only develop a limited number of lines of analysis in a given study. At times they will be restricted in the scope of the data they can collect. At others the capacity for analysing the available data will be limited. In addition researchers may only be able to report a limited proportion of the analyses they have carried out as a result of journal restrictions on the length of papers, as I have already discussed.

CONCLUSIONS

The strength of qualitative analyses, such as that of Morgan and Watkins, lies partly in their ability to generate as many questions as they answer and to point the way forward to future studies. Studies such as this one are best seen, not in isolation, but as a contribution to our cumulative knowledge of how lay people make sense of health and illness and of particular disease diagnoses. Carrying out such studies is far from easy and demands considerable skill and judgement. However, the contribution which they can make to our understanding of patients' responses to both diagnoses and medical advice amply repays the effort involved.

Morgan and Watkins have made use of just one of the numerous methods of qualitative research which are relevant to general practice. Others include observation, group interviews, video and audio-recordings of naturally occurring events, focus groups, and narratives. Helman (1991) gives an extensive list of the available methods, along with examples of how they might be applied in general practice research. As both Helman (1991) and McWhinney (1992) argue, the scope for qualitative methods in general practice is enormous. Again, Helman not only details specific research areas which he believes to be particularly appropriate for qualitative methods, but he also offers copious examples of specific studies which have done so.

Qualitative methods have particular advantages where the aim of a piece of research is to *understand* a situation. We need such methods, as McWhinney argues, 'because some of the key questions in family research will not yield to the current ones' (1992, p. 5). These methods can give access to the meanings which particular events or experiences hold for people. They allow us to explore the cases where generalizations drawn from quantitative studies do not hold (Cronbach 1975). Such an objective is surely appropriate in a discipline, such as general practice, which is primarily concerned with individuals rather than populations.

These qualitative methods minimize the risk that the researcher will impose taken-for-granted assumptions, derived from the medical perspective, upon patients' actions. They offer the excitement of discovering new ways of looking at old problems and the possibility of 'making the implicit explicit' (McWhinney 1992). Unlike much quantitative research, it is possible in qualititative studies to retain much of the context of research findings.

No single method can hope to be appropriate to the diverse needs of a discipline such as general practice. The challenge is to identify both the strengths and weaknesses of the available methods in the context of the objectives of a given piece of research. The final choice of method is perhaps best seen as a trade-off between these strengths and weaknesses. Morgan and Watkins' paper demonstrates that, in such a trade-off, the strengths of a qualitative approach can outweigh its limitations.

REFERENCES

Becker, H. S. and Geer, B. (1957). Participant observation and interviewing: a comparison. *Human Organisation*, **16**, (3), 28–32.

Blaxter, M. (1990). *Health and lifestyles*. Routeledge, London.

Calnan, M. (1987). *Health and illness: the lay perspective*. Tavistock, London.

Cronbach, L. J. (1975). Beyond the two disciplines of scientific psychology, *American Psychologist*, **30**, 116–27.

Glaser, B. G. and Strauss, A. (1967). *The discovery of grounded theory*. Aldine, Chicago.

Guba, E. G. and Lincoln, Y. S. (1981). *Effective evaluation: improving the usefulness of evaluation results through responsive and naturalistic approaches*. Jossey-Bass, San Francisco.

Guba, E. G. and Lincoln, Y. S. (1982). Epistemological and methodological bases of naturalistic inquiry. *Educational Communication and Technology Journal*, **30**, 233–52.

Hammersley, M. (1992). *What's wrong with ethnography?* Routledge, London.

Hammersley, M. (1993). Unit seven. In DEH313 Course Team *Principles of social and educational research*, The Open University: Milton Keynes.

Helman, C. G. (1985). Communication in primary care: the role of patient and practitioner explanatory models. *Social Science and Medicine*, **20**, (9), 923–32.

Helman, C. G. (1991). Research in primary care: the qualitative approach, in *Primary care research: traditional and innovative approaches*. (eds. P. G. Norton, M. Stewart, F. Tudiver, M. J. Bass and E. V. Dunn) Sage Publications, London.

McWhinney, I. R. (1991). Primary care research in the next twenty years. In *Primary care research: traditional and innovative approaches*, (eds. P. G., Norton, M., Stewart, F., Tudiver, M. J., Bass and E. V., Dunn.) pp. 1–11 Sage Publications, Newbury Park.

Morgan, M. and Watkins, C. J. (1988). Managing hypertension: beliefs and responses to medication among cultural groups. *Sociology of Health and Illness*, **10**, 561-78.

Murphy, E. (1992). Lay health concepts and response to medical advice about lifestyle modification. Unpublished PhD thesis, University of Southampton.

Murphy, E. A. and Mattson, B. (1992). Qualitative research and family practice: a marriage made in heaven? *Family Practice*, **9**, (1), 85-91.

Payer, L. (1990). *Medicine and culture*. Victor Gollancz, London.

Pope, C. and Mays, N. (1993). Opening the black box: an encounter in the corridors of health services research. *British Medical Journal*, **306**, 315-18.

Schofield, J. W. (1993). Increasing the generalizability of qualitative research. In *Social, research: philosophy, politics and practice*. (ed. M. Hammersley) Sage Publications: London.

Stimson, G. V. (1974). Obeying doctors' orders – a view from the other side. *Social Science and Medicine*, **8**, 97-105.

Managing hypertension: beliefs and responses to medication among cultural groups

Myfanwy Morgan and C. J. Watkins
Department of Public Health and General Practice, UMDS

Sociology of Health and Illness, **10**, 561–77 (1988)

Abstract

Lay beliefs and responses to antihypertensive therapy were examined through tape recorded interviews held with European respondents, who were predominately English and referred to as 'white', and people born in the West Indies. Respondents were aged 35–55 years, of manual social class and currently being treated for hypertension by their general practitioner. Nearly one-half of the respondents identified stress or worry as a cause of their high blood pressure. Possible reasons for this are considered, including their doctors' explanation of the causes and respondents own understanding of the term 'hypertension'. Respondents were all aware of the importance of controlling their blood pressure. However, whereas the level of adherence to the prescribed medication was high among white patients, less than half of the West Indians were classified as compliers, with many regularly 'leaving off' the drugs. This reflected their particular beliefs and concerns, and was often associated with the use of herbal remedies, forming a continuation of traditional cultural patterns.

INTRODUCTION

Hypertension mainly takes the form of essential hypertension, or an elevation of blood pressure with no known cause. Its significance lies in the increased risks of stroke and coronary heart disease associated with high blood pressure. These risks appear to rise sharply with diastolic levels of 100 mm Hg and over (Rose 1984). There is thus a concern to identify at risk individuals and to control blood pressure levels. However, although a variety of antihypertensive drugs are available which permit blood pressure control in the majority of patients, there are questions of the blood pressure level at which drug therapy should begin. This requires the risks of high blood pressure to be balanced against the costs to the individual of daily medication (Medical Research Council Working Party, 1985; Dollery, 1984). There is also evidence from the United States that the level of compliance with antihypertensive medication is low. About one-third of patients drop out of treatment in the first year, while up to one half of those who do continue in treatment appear to take the tablets irregularly (Weinstein and Stason, 1976; Luscher *et al.*, 1987). As in the USA, a substantial proportion of hypertensive patients in Britain appear to drop out of treatment (Heller and Rose, 1977). However, there is no data on the level of adherence to antihypertensive medication among those who do continue to be treated by their general practitioner.

The question of why patients fail to take their medication as prescribed has been the focus of a large number of studies (Haynes, Taylor and Sackett, 1979). Research has generally adopted a problem-oriented perspective and has been concerned to explain and reduce patient 'defaulting'. Studies of hypertensive patients have identified the perceived efficacy of the anti-hypertensive regime, a high level of anxiety when the condition was first diagnosed, beliefs concerning the necessity of regular medication and the experience of side-effects, as all serving to discriminate between compliers and non-compliers (Nelson *et al.*, 1978; Norman *et al.*, 1985). This approach thus determines the distribution of beliefs which are hypothesized to influence patients' responses to drug therapy. The reasons for non-compliance are then inferred from these associations, with little attempt being made to assess the significance of particular beliefs and concerns for individual behaviours or the variations which may occur across social and cultural groups. In contrast, this study aims to elicit hypertensive patients' own beliefs, concerns and patterns of behaviour, rather than seeking to test specific hypotheses or to determine the distribution of particular beliefs.

The study was conducted in the London borough of Lambeth, an inner London area with a predominately working class population. A notable characteristic of the area is the high proportion of ethnic minorities, who comprise about 23% of the population (Department of Environment, 1983). West Indians form the largest ethnic minority and account for 12% of the population (Department of Environment, 1980). West Indians are also of particular interest as they appear to have a high prevalence of hypertension compared with 'white' and Asian populations (Cruickshank, Beevers and Osbourne, 1980). This study thus compares the beliefs and responses to hypertension of a small group of 'white' and West Indian patients of similar socio-economic status.

METHOD

Fifteen general practices were approached to identify patients currently being treated for hypertension. Six practices kept hypertension registers and in the other nine general practitioners were asked to provide a list of their 'white' and West Indian hypertensive patients. It was intended to select for interview equal numbers of men and women who were aged 35–55, of manual social class, and who were not currently receiving medication for any other chronic condition, with the aim of reducing the effect of these variables. General practitioners were therefore asked to identify their hypertensive patients who satisfied these criteria. Further information, including the most recent blood pressure measurement and medication prescribed, was extracted by the researcher from the patients' case notes.

From an initial group of 74 patients, 65 were contacted, of whom all but three agreed to be interviewed. Two respondents were omitted from the analysis because on interview they were classified as being of non-manual social class. The study is thus based on interviews held with 30 white patients and 30 West Indian patients, with both groups equally divided between men and women. The selection of respondents from general practice lists of hypertensive patients meant that the sample excluded people who may have dropped out of treatment completely, as well as those with undiagnosed hypertension.

The interviews were conducted in the respondents' own home and generally lasted just under one hour. All interviews were-tape recorded and subsequently transcribed. They were based on a schedule of open-ended questions and conducted in a relaxed, conversational manner, with the interviewer trying to adopt the role of a 'friendly listener'. Respondents were informed that their general practitioner had given their name as a patient with high blood pressure but assured of the confidentiality of their responses.

The interview was introduced with easy, non-probing questions, while all questions were phrased in such a way as to try and ensure that respondents did not feel there was an expected or acceptable answer and instead were encouraged to talk freely about their own beliefs, concerns and practices. For example, questions concerning worries about having high blood pressure were prefaced by acknowledging that some people get quite worried about having high blood pressure, and then asking whether having high blood pressure worried them, whether they thought about it often or just occasionally, and what they worried about, etc. Similarly, questions about taking their prescribed medication were introduced by stating that some people often forgot to take their drugs, and then asking respondents whether they took their tablets every day or just occasionally, if they had difficulty remembering, and how often they forgot to take the tablets, etc. In addition, more detailed questions were asked about drug-taking during the previous week. Although the interviewer did not ask to see or count their tablets, respondents who claimed to be taking anti-hypertensive drugs regularly often showed them to the interviewer and described exactly when they took them and their routines for remembering, and appeared to be genuine compliers.

CHARACTERISTICS OF RESPONDENTS

The West Indian respondents were born overseas, with all but three coming from Jamaica. They had come to England as young adults between 1958 to 1964 in response to the employment opportunities and recruitment drives. The respondents referred to for brevity as 'white' were mostly born in England. However, four came from southern Ireland and one from Spain. About three-quarters of all respondents had been living in Lambeth for over 15 years and none less than 5 years, while one third of the white respondents had lived in Lambeth for their entire life.

Respondents were all aged 35–55 years. However, they were mainly in the upper part of the age range, with only one quarter aged under 45 years. There was no difference in the age distribution of West Indian and 'white' respondents, or between men and women. The occupational distribution of the two ethnic groups was also very similar. Altogether 24 of the 30 men were currently employed. Common occupations were cleaner, caretaker, railway worker and general labourer, with only small numbers engaged in skilled trades. Twenty-five of the 30 women were employed either full or part-time, mainly as clerical workers, cleaners, cashiers or shop assistants. In terms of housing, 21 of the 30 West Indians and 28 'white' respondents were living in rented accommodation, which mainly consisted of council flats on large estates. Nine West Indians and two 'white' respondents were owner occupiers.

Patients were selected for interview only if they had been diagnosed as hypertensive for at least one year, to ensure that stable patterns had been established. Just under half of both the white and West Indian respondents had been diagnosed as hypertensive more than 5 years ago and four people were first diagnosed over 10 years ago.

The latest blood pressure reading recorded in the patients' case notes gave some indication of their present blood pressure status and indicated that the West Indians may have been less well controlled. Only 10 of the 30 white respondents had a diastolic BP of over 100 mm Hg, whereas 16 West Indians had a diastolic BP of 100 mm Hg or over and included three with a diastolic BP of 120 mm Hg or over.

BELIEFS ABOUT HIGH BLOOD PRESSURE

Twenty-seven of the 60 respondents identified stress, worry or tension as a cause of their high blood pressure (Table 1). For six respondents this was because they viewed themselves as being a 'worrier':

I think that I get worked up very easily. Just worry about things and live on my nerves. It could be that.

More commonly people identified specific sources of stress. Family problems formed a major source and included the illness or death of a family member, and problems with a spouse or child. Another cause of stress frequently mentioned by men was unemployment or pressures at work. One man currently employed as a driver explained, when asked about the causes of his high blood pressure:

I was under a lot of pressure at the time, being out of work for 9 months. I couldn't get a job. I applied for plenty of jobs and knew plenty of people at the Council and various other places and yet I couldn't get one. That helped build it up (high BP). That could have been the reason I got this blood pressure problem.

Similarly, a West Indian factory foreman explained:

With the job that I'm doing, its pretty hectic because you have to supervise your own people. You get a lot of hassle from that. You try to keep calm but it's not easy if you are supervising 30 people and everybody calling you at the one time. It's stressful. I think that helped raise my pressure.

Other frequently mentioned causes were familial and hereditary factors, diet and overweight. The dietary factors mentioned by both groups were too much salt and fat, with too much starchy food, such as yam, also being identified by West Indian respondents.

The main difference between ethnic groups was that 13 white respondents but

Table 1 *Respondents' perceived causes of their high blood pressure*

Causes mentioned	White respondents (N = 30)	West Indian respondents (N = 30)
Tension, worry, stress	14	13
Familial, hereditary	3	4
Diet, smoking, alcohol	3	7
Overweight	3	1
Other causes	2	2
No. identifying cause	17	24
Mean no. causes reported	1.5	1.1

Table 2 *Respondents understanding of the term 'hypertension'*

Meaning of hypertension	White respondents (N = 30)	West Indian respondents (N = 30)
Same as high blood pressure	14	16
Stress, worry, tension etc.	5	6
Don't know meaning	11	8

only 6 West Indians were categorised as unable to suggest the cause of their high blood pressure (Table 1). During the interview the term high blood pressure was generally used. However respondents were asked if they had heard of the word 'hypertension' and if they knew what this meant. Nineteen people did not have any idea of the meaning, although most said they had heard the term before (Table 2). Of the other 41 respondents, 30 regarded hypertension as being just a different name for high blood pressure, although only one half of this group had identified stress, worry or tension as a cause of their high blood pressure, suggesting that they were not necessarily equating high blood pressure with being 'hyper-tense'. The other eleven respondents described hypertension specifically in terms of 'having lots of worries', being 'highly stressed' or 'getting worked up'. However, most of this latter group seemed to regard hypertension and high blood pressure as separate and quite distinct conditions, with only the latter applying to them. Only three people viewed these conditions as linked, with hypertension described as '. . . nervous problems or troubles of the mind. Things that bring it on (high blood pressure), and build it up'. As Table 2 shows, there was no difference between white and West Indian respondents in their understanding of the term 'hypertension'.

When asked if their doctor had told them anything about the causes of their high blood pressure, 35 respondents said that their doctor had not told them anything, while a further six said that their doctor had explained that no-one knew the causes of high blood pressure. Thus for about two-thirds of respondents their doctor does not appear to have explained or suggested the causes of their high blood pressure. However, 13 respondents (7 white and 6 West Indians), said that their doctor had asked if they had any worries or suggested that their stressful circumstances might have caused their high blood pressure, while another 6 respondents said that their doctor seemed to think that they might have inherited a tendency to high blood pressure. There was again no difference between white and West Indian respondents in terms of what their doctor had told them about the causes of their high blood pressure.

With regard to the experience of high blood pressure, 35 of the 60 respondents felt they could tell if their blood pressure was better or worse (Table 3). The most frequent indication that their blood pressure had gone up was the experience of pains or 'sensations' in the head. Sometimes this was just described as a headache. However frequently it was described as different from an ordinary headache:

Just there (indicates forehead). It's a mucky feeling you have there. It don't pain you, it s just funny. You have a funny feeling there and then you know its high.

Table 3 *Symptoms perceived as indicating a rise in blood pressure*

Symptoms	White respondents (N = 30)	West Indian respondents (N = 30)
Pains or sensations in head	7	15
Weakness, tiredness	2	4
Eye problems	3	3
Dizziness	2	1
Feeling hot	3	1
No. noticing changes in BP	15	20
Mean no. symptoms reported	1.1	1.2

Yes, I can tell my pressure is up when I get this fuzzy feeling at the top of my head . . . not a headache, it's a fuzzy feeling, a light headed feeling.

These feelings and pains in their head were most common among the West Indian respondents, and were generally attributed to worry, stress and tension, or to rushing around too much.

Respondents were all aware that you can die if you have high blood pressure which is not controlled, and 56 of the 60 respondents identified a heart attack or stroke as the likely causes of death. When asked if they worried about having high blood pressure, 28 of the 60 respondents (17 whites and 11 West Indians) acknowledged that they did worry, although most explained that this was not a major or constant worry. An important trigger causing people to worry was their own feeling that their blood pressure was high or being told this by the doctor.

People who claimed not to worry about their high blood pressure often explained that there was no need to worry. This was often because 'the doctor can get your blood pressure down', or they thought that their blood pressure was under control. In addition, several white respondents explained that 'there's a lot more worse things in life to cope with than having blood pressure'. The West Indians more commonly emphasized the normality of high blood pressure in explaining why they did not worry about it:

I knew a lot of people who have high blood pressure, so I wasn't worried at all.

Table 4 *Use of anti-hypertensive drugs*

Use of drugs	White respondents (N = 30)	West Indian respondents (N = 30)
Takes as prescribed	26	12
Often forgets	1	2
Takes reduced dosage	–	2
Regularly 'leaves off'	1	14
Recently taken off medication by GP	2	–

No, I didn't (feel worried) because you see my mum had it and also my mum's sister and brother as well. It's in my family, so I'm not really worried.

Both white and West Indian respondents also frequently commented that they would prefer to die of a heart attack than from cancer. This was because they viewed a heart attack as leading to a quick death without much pain, compared with the slow, painful death they associated with cancer, and reflects the general image of cancer as the most dreaded disease (Sontag, 1979).

The main difference between cultural groups were therefore that West Indians more frequently identified a cause of their high blood pressure and were also rather less likely to report that they worried about having high blood pressure. However, all respondents were aware of the importance of controlling their blood pressure.

RESPONSES TO DRUG THERAPY

Details of the prescribed drugs recorded in the patients' case notes showed that the three most common drug regimes were a combination of diuretic and beta blocker (25 patients), followed by a beta blocker only (13 patients) and diuretic only (8 patients), with 14 patients receiving other types of medication. None of the individual drugs prescribed was recognised as producing a high incidence of side effects. There was also no difference in the drugs prescribed for the West Indian and 'white' respondents.

Respondents were asked about their experience of side-effects. One third (9 'whites' and 11 West Indians) said that initially they had experienced considerable problems with side-effects; mainly feelings of dizziness, sickness and tiredness, and impotence among men. As a result their tablets had been changed. This appeared largely to resolve the problem, with few side effects being reported with their current medication.

In response to questions about taking their drugs, 26 of the 30 'white' respondents said they took their tablets regularly as prescribed. Of the other four respondents, two said that their general practitioner had taken them off the drugs at their last visit as their blood pressure was controlled, while just two 'white' respondents were classified as non-compliers. In contrast, only 12 of the 30 West Indians said they were taking their medication as prescribed (Table 4). Compliers all appeared to attach considerable importance to following their doctors' instructions and taking the tablets regularly, frequently emphasizing that you should 'do what the doctor says, and then you can't go wrong'. Several of those who took their medication regularly acknowledged that they might forget the occasional tablet. However, only three people said they often forgot the tablets. Remembering was generally achieved through establishing a routine for taking the tablets, while additional supplies might be kept in other places, such as at work. Respondents also often referred to the importance of their spouse in helping to ensure that they remembered to take the tablets.

Two West Indians were taking a reduced dosage because of their experience of side effects with the prescribed dosage. However, the major difference between the 'white' and West Indian respondents was the large numbers of West Indians who frequently 'left off' the tablets. This involved either not taking the drugs for a few days each week, or not taking them for a week or two at a time, or even for a couple of months. This practice of 'leaving off' the tablets occurred among both men and women and at all blood pressure levels. Currently, eight of the 14 West Indians with a diastolic BP of under 100 mm Hg claimed to 'leave off' the drugs, as did six of the 16 with a diastolic BP of 100 mm Hg or over. People who regularly 'left off' their drugs were all aware that their doctor expected them to take the tablets every day. This behaviour was therefore a deliberate action: it could not be attributed to problems of understanding or forgetting.

One reason for 'leaving off' the tablets mentioned by four West Indian men was the danger of mixing the tablets with alcohol. If they were planning to drink spirits, often over the weekend, they would therefore not take the tablets for a few days. A more common·reason for leaving off the tablets by both men and women was their concern about the possible harmful effects of long term drug therapy and fears of becoming 'addicted' to the drugs. For example:

Sometimes I remember and sometimes I don't because I don't want to build my hopes on tablets. I don't want to become an addict.

It sometime be 5 or 6 months and no tablets go down my throat. I don't use them. I see too many people on these tablets having side effects, so I try to get away from tablets. I really must be gone bad to take tablets because I don't like tablets.

Linked with this concern about the harmful effects of the drugs was frequently a questioning of the need to take the drugs if they were feeling all right. As two respondents explained:

I find it difficult if I'm feeling fine for two or three weeks to keep taking a pill. At the end of the day is that pill going to make me better or worse? It's for making you better, right. So once you feel better, why continue taking it? If I have a headache I take asprin. Your headache goes. You don't take the pills if you don't have a headache.

If I was feeling fine for a day or two then maybe I would miss a day or two. If that's cheating on myself, then thats it . . . apart from helping the blood pressure they may cause something else, and also addiction as well.

Although respondents frequently stopped taking the drugs for a time, they appeared to be engaged in a continuous process of monitoring their condition and the effects of taking or 'leaving off' the tablets. An important trigger to taking tablets more regularly was being told by their doctor that their blood pressure had gone up. Some respondents who were now taking their drugs as prescribed were doing so in response to this information. However, sometimes a visit to the doctor confused the picture as these two respondents explained:

Sometimes I go there (to the doctor) and they check the pressure and say it is high, but I've been taking the tablets everyday. Next time I go they say that it is not too bad, so I'm not sure what's happening.

Sometimes even though I have been taking them (the tablets) everyday the pressure is still up, and then I stop from taking them and the pressure is down. Well it don't make any sense.

(Did you tell your doctor that?)

No, I don't tell her anything.

When they felt their own blood pressure was high, and especially if they experienced pains in their head, both ethnic groups responded by resting and not rushing around as much. The West Indians who frequently left off the tablets also explained that in this situation:

... you probably take the tablet, and use the tablet not as a blood pressure tablet but as a tranquilliser because you feel you are in a state.

When I feel in my head that I have a headache then I automatically take a couple (of tablets) and then it goes away.

ALTERNATIVE REMEDIES

All respondents were asked whether they took herbal, homeopathic or other remedies for their high blood pressure. Just one white respondent and 17 of the 30 West Indian respondents said they regularly took herbal remedies to promote their general health and help their blood pressure. These herbal remedies were generally referred to as 'bitters' and can be purchased locally. One variety sold in a market in Brixton is described as 'A Blood Toner' suitable for high blood pressure, diabetics, billiousness, loss of appetite and general debility, and lists eight herbs and roots as the ingredients. Alternatively, a herbal tea is made, often from fresh or dried sorocee.

The use of bitters is a tradition brought over from Jamaica where many different herbs are used as a tonic to promote health and as a cure for illness. As one respondent explained:

There are lots of things you get in the West Indies that you can't get over here. You can go up the mountains and pick herbs, charney root and bloodweiss. You boil them together with watergrass.

(Is that what you would use for your blood pressure if you were still living in the West Indies?)

Yes. All those things bring your pressure right down.

Herbal remedies were often taken in addition to the prescribed drugs. However, patients who decided to 'leave off' their drugs for a while might then rely on the herbal remedies and monitor the effects on their blood pressure. If after a period of only taking the herbal bitters their general practitioner said their blood pressure was 'fine', this was taken to indicate that the bitters were doing them good. However, if they were told their blood pressure had gone up they would then start taking their prescribed medication again. Most respondents said they did not tell their general practitioner that they took herbal remedies or that they 'left off' their drugs. This was because they thought their doctor would disapprove and think them stupid, as they did not think general practitioners understood West Indian habits.

Respondents did not appear to be taking any other remedies for their high blood pressure, or engaging in yoga, relaxation classes or other activities. However, whereas the white respondents relied exclusively on National Health Service general practitioners, one third of the West Indians occasionally consulted a doctor as a private patient for a second opinion. This practice again reflected their cultural traditions, as one respondent explained:

In the West Indies we have a lot of private doctors. We have always had private doctors. We pay them and expect a service from them, which they give. You can call him anytime, day or night . . . It's no good going to a doctor and he starts to write a prescription out, like he is working in a factory. There's no time available. . . . Some of us realise this, so that's why we go and get another doctor.

Five West Indian respondents had consulted a general practitioner privately about their blood pressure when they felt very worried about it being high. This resulted in a confirmation of the diagnosis and either a course of tablets being given or reassurance about the appropriateness of those already prescribed. When different tablets were prescribed their effects were monitored by the respondent and subsequently discussed with the general practitioner with whom they were registered.

DISCUSSION

A notable feature of both the white and West Indian respondents' beliefs about their high blood pressure was the identification of worry, stress and tension as a cause. This was mentioned by 27 of the 60 respondents and formed the most common single cause identified. A similar emphasis on worry, stress and tension as the cause of hypertension was found in Blumhagen's (1980) study of men attending a hypertensive clinic in Seattle, USA. The recent *Health and Lifestyle Survey* based on a national sample of over 9,000 adults aged 18 and over in England, Wales and Scotland also showed that 53% of people with treated hypertension identified tension, worry and stress as a cause of high blood pressure, as did 55% of untreated normotensives (Blaxter, 1987). The lay identification of stress as a cause of hypertension thus appears to be fundamental and cross-cultural. However, although it is accepted that stress may initiate a sudden and transient elevation in blood pressure and sustained elevations have been produced in animal studies, there is at present no conclusive scientific evidence that stress can produce sustained blood pressure changes in man (Ostfeld and Shekelle, 1967; Hart, 1980). However, as Marmot (1985) notes, there is a plausible biological explanation of how such a long term effect could be produced. Indirect evidence of the contribution of stress has also been provided by trials of stress management techniques, which have successfully lowered blood pressure levels and maintained this difference compared with a control group over a four year follow-up period (Patel *et al.*, 1985). However, problems of defining and measuring stress have resulted in relatively little attention being paid to its role as an epidemiological risk factor.

Although the lay perception of stress as a cause of high blood pressure may have some as yet unproven basis in terms of the pathophysiological mechanisms underlying hypertension, there is still the question of the sources of these lay beliefs. One explanation suggested by Blumhagen (1980) is that lay people tend to interpret hypertension as 'hyper-tension' or excessive tenseness, and often regard high blood pressure as one of possibly many symptoms of 'hyper-tension'. In this study, 11 of the 60 respondents interpreted hypertension to refer to stress, worry and tension. However, only three people regarded hypertension as leading to high blood pressure, whereas the other 8 respondents viewed hypertension as an entirely separate condition from high blood pressure. In addition, although

30 people regarded hypertension as being the same as high blood pressure, they did not necessarily identify worry, stress or tension as a cause of their high blood pressure. Thus although patients' misunderstanding of the term 'hypertension' may in a few cases have contributed to a belief in stress or tension as a cause of high blood pressure, this does not appear to be responsible for the widespread emphasis on worry, stress and tension as a cause among the present group of general practice patients.

A second possible reason for the lay emphasis on stress and tension is that doctors may sometimes suggest that stress forms a cause of high blood pressure by their questions to patients. Among the present group of respondents, nineteen said that their doctor had discussed specific causes, while 13 respondents said that their doctor had mentioned stress and tension. This indicates that some general practitioners may accept the potential role of stress in increasing risks of hypertension in the operational model they employ on a day-to-day basis. In other cases, although not personally accepting this explanation, they may not try to alter lay beliefs and dissuade a person who believes their high blood pressure was caused by stress, which may in turn serve to perpetuate and reinforce this lay belief.

Doctors' and lay people's explanatory models can thus be viewed as influencing each other as part of a dynamic process, for just as professional beliefs and practices influence the lay culture, so the operational models employed by general practitioners may be shaped both by medical science and by their participation and acceptance of the lay culture. In particular, despite the lack of scientific evidence as to the role of stress in producing a sustained elevation in blood pressure, clinical textbooks frequently list stress, together with heredity and salt intake as the causes of hypertension (Houston *et al.*, 1982).

Although both the white and West Indian respondents were aware of the importance of controlling their blood pressure, they differed in their responses to anti-hypertensive medication. Almost all the white respondents said they took the tablets as prescribed, compared with only 12 of the 30 West Indian respondents. Respondents who claimed to be taking their tablets regularly appeared to be genuine 'compliers', although they frequently acknowledged that they sometimes missed the occasional tablet, especially if their routines were disrupted.

An important form of non-compliance reported by one-third of the West Indian respondents was to regularly *'leave off'* their anti-hypertensive medication. This was generally explained as being due to their concern about the long-term harmful effects of drug taking and a dislike of being dependent on drugs. They also frequently questioned the need to take the tablets if they felt all right, despite being aware of their doctor's advice and expectations. This regular 'leaving off' of the prescribed medication by West Indian respondents was thus a response to their evaluation of the necessity and costs of drug taking, rather than reflecting problems of side effects or a failure to understand the drug regime or to appreciate the importance of controlling their blood pressure. The dislike and partial rejection of long-term drug therapy by this group of first generation immigrants was frequently associated with the use of herbal remedies, taken either as well as, or as an alternative to, the prescribed medication. In addition, they frequently sought second opinions from a private general practitioner, reflecting their previous experience and expectations as a consumer in a fee-for-service system. A recent survey of responses to illness and the use of health services carried out in Jamaica identified important similarities with the present study (MacCormack, 1985). For example, when asked what people can do about high blood pressure, 58% of the sample thought you should go to the doctor and take medicine but this was usually combined with other self-help remedies. Included among these were the use of herbal remedies mentioned by 27% of the sample, with over fifteen different types of herbal tea being identified as helpful for high blood pressure. Thus although the present group of West Indian respondents had come to Britain as young adults and have lived in this country for over 20 years, their responses to high blood pressure, and to illness and medical care more generally, reflected elements of their cultural traditions and practices. This raises questions of the extent to which the beliefs and health practices of this group of first generation West Indian immigrants have been transmitted to their children who were born and brought up in this country. There are also questions of the effects of the regular 'leaving off' of the prescribed medication on blood pressure levels. West Indians appear to have a relatively high incidence of hypertension, and differences in salt intake or salt retention have been identified as possible risk factors (Grell, 1983). However, the higher blood pressure levels of hypertensive blacks than hypertensive whites found in a study by Sever *et al.*, (1978) and the larger numbers of West Indian patients with uncontrolled blood pressure among the present group of respondents may also be associated with the observed variations in adherence to the prescribed medication, as well as with possible differences in rates of drop-out from treatment.

The use of folk remedies in Britain is probably currently most common among ethnic minorities, and especially among immigrants from rural areas with less access to modern scientific medicine. In the U.K. a generational difference in self-help responses has been identified by Blaxter and Paterson (1982) in a study of disadvantaged families in Aberdeen. The grandmothers' generation frequently used folk remedies for a range of illnesses, whereas their daughters relied more heavily on proprietary medicines brought from the chemist. In recent years there has been a trend towards the increasing use of homeopathy, yoga, acupuncture and other healing systems, which form a large and growing 'hidden health care system', and is often sought when conventional medicine is unable to provide an effective cure (Helman, 1987). However, recourse to these systems may be more common among middle class than working class people. Responses to illness are affected by a complex interplay between class and culture.

Department of Community Medicine and General Practice
United Medical and Dental Schools
St. Thomas's Hospital
London SE1

ACKNOWLEDGEMENTS

We would like to thank Judy Martin who conducted the interviews and Mary Evans who assisted with the administration. We have also benefited from helpful discussions with *Mildred Blaxter, Linda Jarrett and David Morrell*. The research was supported by the Department of Health and Social Security.

REFERENCES

Blaxter, M. and Paterson, E. (1982) *Mothers and Daughters: a three generational study of health attitudes and behaviour*. London: Heinemann Educational Books.

Blaxter, M. (1987) Beliefs about the causes of ill health. In Cox, D. *et al.*, *The Health and Lifestyle Survey*. London: Health Promotion Research Trust.

Blumhagen, D. (1980) Hyper-tension: a folk illness with a medical name. *Culture, Medicine and Psychiatry*, **4**, 197–227.

Cruickshank, J. K., Beevers, D.G., Osbourne, V. L. *et al.* (1980) Heart attack, stroke, diabetes and hypertension in West Indians, Asians and Whites in Birmingham, England. *British Medical Journal*, **281**, 1108.

Department of the Environment (1980) *National Dwelling and Housing Survey*, London: HMSO.

Department of Environment (1983) *Census 1981, Information Note No. 2* London.

Dollery, C. T. (1984) Management of hypertension: risk-benefit ratio. *Journal of Hypertension 2* (supp 2), 9–12.

Garrity, T. (1981) Medical compliance and the clinician-patient relationship: a review. *Social Science and Medicine* 15E, 216–22.

Grell, G. A. C. (1983) Hypertension in the West Indies. *Postgraduate Medical Journal* **59**, 616–21.

Hart, J. T. (1981) *Hypertension*. London: Churchill-Livingstone.

Haynes, R. B., Taylor, D. N. and Sackett, D. L. (1979) *Compliance in Health Care*. Baltimore: The Johns Hopkins University Press.

Houston, J. C., Joiner, C.L. and Trounce, J. R. (1982) *A Short Textbook Medicine* (7th ed) London, Hodder and Stoughton.

Helman, C. G. (1987) General practice and the hidden health care system. *Journal of Royal Society of Medicine*, **80**, 138–40.

4 Fitting the method to the question; the quantitative or qualitative approach?

Roisin Pill

INTRODUCTION

The argument presented here is that the choice of method involves more than a simple decision about how to collect data. It carries with it underlying assumptions about what constitutes an appropriate question for investigation and indeed how findings are to be judged and interpreted. The paper chosen for discussion is used to illustrate this proposition. It is argued that while the survey method may provide quantifiable data to answer specific questions, it is not necessarily the best way of uncovering the range and depth of peoples feelings and opinions. Another approach using qualitative methods might have addressed the question of patients' views on health promotion in a different way and produced different results. This is not to suggest that one approach is necessarily *better* than the other: the intention is to point up the fact that they reflect different research traditions and to sensitize the general practitioner to the implications of opting for one rather than the other.

THE CONTEXT IN WHICH THE PAPER WAS WRITTEN

The identification of prevention as one of the key responsibilities of the general practitioner is not a particularly recent phenomenon and indeed the implications of this role have been a recurring theme in both the general practice literature and the publications on wider issues of prevention and health promotion. By the 1970s changes in national and international health policy gave prevention a new prominence (DHSS 1976, 1977; WHO 1978). This shift in approach was associated with a growing awareness of epidemiological research into the aetiology of major acute and chronic diseases where attempts had been made to quantify risk in relation to different groups within a population. The concepts of 'lifestyle' and 'risk factor' gained wider currency and by 1983, Doll and Peto were able to sum up their view of the potential for prevention by setting out the key evidence for links between mortality and morbidity and social, environmental, and cultural risk factors within population.

In response to these trends the Royal College of General Practitioners set up a series of working parties to examine the issues in the context of general practice and to indicate what the scope should be (RCGP 1981 *a, b, c, d*, 1982, 1983). There were already a number of publications stressing the particular features of general practice that made it a natural setting for preventive medicine and health promotion (Stott and Davis 1979; Smail 1982; Stott 1983) and others, mainly by social scientists, that were beginning to examine these claims more closely in the light of empirical evidence (Boulton and Williams 1983; Calnan and Johnson 1983). Boulton and Williams, for example, pointed out that although a relatively large proportion of consultations in their study provided opportunities for problem-related education these were not often taken up. There were also a number of studies from general practitioners themselves describing interventions which were considered to be effective (Handel 1973; Porter and McCullough 1972; Jamrozik and Fowler 1982). The most frequently quoted research study, and the one that was most rigorous in its method, focused on the evaluation of the impact of doctors' advice about the dangers of smoking (Russell *et al.* 1979). This concluded that general practitioners can have significant impact, if only a small one, on smoking behaviour.

REASONS FOR SELECTION

This was the context in which the paper chosen for discussion here (Wallace and Haines 1984) appeared in 1984. Several considerations influenced my choice; as noted above, the topic of health promotion in the consultation was under debate at that particular time and has gone on being debated until the present. Indeed Professor Stott and I have contributed to that debate through publishing our own research findings and therefore I must declare a personal interest in the topic (Pill and Stott 1987; Pill *et al.* 1989; Stott and Pill 1990*a*).

However, the most important reason for selecting this particular paper for critical analysis is, in my view, the nature of the research question posed. Instead of rehearsing the claims of general practice to undertake health promotion, describing what was being done or attempting to evaluate the results, the authors took the imaginative step of wondering what the *patients*, not the doctors, thought about the increasing trend for family doctors to undertake preventive activities and to give advice designed to promote changes in certain behaviours designated as 'risk factors'. They deserve credit for recognizing that the opinions, beliefs, and attitudes of the recipients of health promotion merited attention. The question posed therefore, was an interesting and important area for research and there are a number of ways it might have been tackled.

The approach the authors adopted was the cross-sectional survey, characteristic of so much general practice research, using a self administered postal questionnaire. This method is typical of the traditional research paradigm drawn from the natural sciences. It relies on the manipulation of quantitative data and deductive reasoning and great stress is laid on sampling procedures in order to ensure the generalizability of any findings. Considerable emphasis is also laid on standardization of data collection procedures to minimize the possibility of unreliable measurement and to allow replication by others.

The point to be noted here is that by choosing to collect the data using quantitative survey techniques the authors opted for a particular tradition of carrying out research. This implies that the reliability and validity of the study can be assessed in certain ways and that the data will be collected, analysed, and presented according to specific rules.

CRITIQUE OF THE METHOD USED

'Any given design is necessarily an interplay of measures, practicalities, methodological choices, creativity and personal judgement by the people involved'
(Patton 1987 p. 9)

The main aim of the paper was to 'determine whether patients think that it is appropriate for general practitioners to be interested in problems relating to poor areas of lifestyle'. The implicit hypothesis in fact being tested was that the majority of patients *do* feel that it is appropriate for their general practitioners to undertake health promotion in the consultation. The forced-choice format of the questions produced quantifiable 'facts' and enabled the authors to conclude confidently that this had been demonstrated and to argue that there may be considerable scope for increasing the emphasis on health promotion in general practice. It is interesting to speculate what would have happened if the authors had not received such a mandate from these respondents. Would a much lower percentage giving consent have been interpreted as posing a problem, perhaps one needing appropriate educational intervention?

Self-administered structured questionnaires are an efficient and cost-effective way of collecting data that lend themselves to statistical analysis. In this case the respondents were patients registered with a general practice and the options were to sample from a number of practices serving areas with a range of socio-economic and ethnic populations or to concentrate on one or two practices only. The authors decided on the latter strategy and surveyed 3452 people, the majority of the adult population registered with two practices serving a council housing estate. Their choice of study populations is open to criticism on the grounds that the respondents were therefore unlikely to span the social class spectrum and that therefore the generalizability of their findings could be questioned. This issue is not acknowledged in the discussion section of the paper. A minor query also arises over the decision to exclude patients over 70 years of age from the survey. It is unclear whether the numbers involved were so small that they could safely be ignored or whether the authors felt that health promotion was largely irrelevant for this age group and therefore their responses were of no interest.

Having decided to survey all eligible adults registered the detailed methodology is clearly explained. The distinction is made between the nominal population to be surveyed drawn from the age–sex register and the true population base of patients still alive and living in the practice area. The authors are to be congratulated on the high response rate achieved for a postal questionnaire (72 per cent). Their persistence in following-up non-responders by further mailed reminders and then tagging the notes and the personal involvement of the general practitioner must have convinced the patients of the importance attached to their replies. Against that there is always the possibility that some patients may have felt pressured by the extent of follow-up to give the reply they assumed would please the doctors.

Although the proportion of non-responders (28 per cent) is lower than in many postal surveys still there is no data on over one-quarter of the surveyed population and the question of possible bias introduced by the omission of this group is not considered in the discussion. We are told that men and younger age groups are under-represented and these are the categories that many medical experts consider have the most to gain from modifying their lifestyle habits.

Thus, like virtually any survey, this one is open to criticism on the grounds of generalizability but within its limits it is a good example of its type. (The authors subsequently replicated the questions on a sample of 47 general practices and produced broadly similar results Wallace *et al.* 1987.)

There were other subsidiary findings that were of interest and hinted at greater complexity. For example it is clear that health promotion was not seen as equally appropriate for all four behaviours considered although it was not possible to determine why. Moreover, there was a discrepancy between the doctors' accounts of their health promotion activity and the patients' perceptions and a suggestion also that men received more input from their doctors than women. These findings are only mentioned. The main conclusions were the positive mandate given for health promotion and the plea for carefully-designed trial intervention (to match the work already done for smoking) in order to evaluate the effectiveness of the general practitioner in the other main areas of risk.

SUBSEQUENT RESEARCH ON HEALTH PROMOTION
IN THE CONSULTATION

The issue of health promotion in the consultation has continued as one of the recurring themes in general practice publications and assumed even greater prominence since, under the new contract, it was enshrined in the duties of every general practitioner. However, the views of patients have not been the focus of published research recently, presumably because there was no longer any doubt that health promotion was acceptable and indeed demanded by patients.

Interest shifted to assessing the effectiveness of interventions in general practice (Jamrozik *et al.* 1984; Russell *et al.* 1987; Wallace 1988) and evaluating the effectiveness of other members of the primary health care team (Sanders *et al.* 1989).

There has also been growing awareness of the discrepancy between medical rhetoric and patient reality hinted at in the findings of the paper under discussion. On the one hand surveys of general practitioners suggest that doctors are very positive about health promotion (Coulter and Schofield 1991) while, on the other, surveys of patients reveal that there is room for more appropriate activity (Wallace *et al.* 1987; Silagy *et al.* 1992). The important research carried out by social scientists for the Health Education Council (Tuckett *et al.* 1985) has provided empirical data on the amount of preventive activity that is actually undertaken in the consultation.

Tuckett and his colleagues set out to study the extent to which there was an exchange of ideas between the doctor and patient by establishing whether the doctor seemed to take a view on each of four topics: diagnostic-significance, treatment-action, preventive-action, and implications. (For each topic a judgement was made as to whether a doctor had said something to indicate his view, done something to enable one to infer that view, or whether there was no indication of what his view was at all.) A key finding was that, while the doctors seemed to have a view about the first two topics in nearly all the consultations sampled, they were much less likely to indicate whether prevention was relevant (only 31 per cent of consultations). The sample consisted of 405 consultations carried out by 16 doctors. It is worth noting that this finding relates to doctors described as dedicated, hard working, and either keen to improve or responsible for teaching consultation skills and highlights the gap between attitude and actual behaviour.

Williams and Boulton (1988) went on to demonstrate the variety of general practitioners' concepts of prevention in a detailed ethnographic study of 34 experienced practitioners, all senior members of the groups responsible for postgraduate and continuing education in two regions. The interviewers explored their views on the potential for prevention in general practice and the differing constraints they perceived in putting this into effect. What was striking about this study was that while the majority of informants were familiar with the recent debates about prevention they interpreted their role in quite different ways; four different groups were identified by the authors which differed on the basis of moral considerations as well as their perceptions of medical knowledge and of the patients. Clearly the amount and content of any health promotion received by a patient could vary greatly depending on the orientation of the doctor involved, particularly the interest and priority individual doctors attach to their preventive function.

This work is an example of an alternative research approach using qualitative methods. As Murphy has pointed out both in collaboration with Mattson (1992) and in Chapter 3 of this book such methods are related to a different philosophical tradition, are best equipped to address different research questions, and often provide fundamentally different types of answer. Such qualitative studies aim for explanation and understanding rather than prediction and do this by exploring the perspective of the actor (Patton 1987). Unlike the dominant research paradigm generalizability is not the prime aim; the details of the particular context are regarded as crucial to understanding the respondents' experiences and while it is possible that insights/concepts/analysis developed for a particular study may be applicable elsewhere it is not claimed that this will necessarily occur.

While Williams and Boulton (1988) developed their analysis to reach greater understanding of why doctors were often not carrying out health promotion despite the apparent widespread acceptance of this role, could such an approach also be applied to patients?

This broad research question, about which there was very little empirical data, was in fact redefined as a specific hypothesis tested by a survey using structured questions to generate quantifiable data. The hypothesis was designed to reassure doctors that what was being claimed as a key feature of modern general practice was approved by the patients, the recipients of all this health promotion activity.

Another option would have been to explore in much more detail exactly what experiences patients had had of health promotion, their attitudes to their general practitioners, and the circumstances under which they felt it was or was not appropriate for the doctor to bring up such topics in the consultation. The aim would be to avoid imposing perceived ideas on the respondent, but instead to draw them out and understand their perspective by using less structured and more open questions.

Such qualitative methods typically produce rich complex data; 'they have the potential to describe phenomena in all their complexity and ambiguity with appropriate consideration of context and attention to the *meaning* of events and experiences for participants' (Zyzanski *et al.* 1992). Such an approach for this topic also implies very different assumptions about the relative role of doctor and patient and what is going on in the consultation. Exploration of patients' views places a value on their opinions, attitudes, and beliefs, and recognizes that their contribution to the consultation needs to be understood. The emphasis is on explaining and understanding the process and the constraints on interaction in the particular context of the consultation; such data can illuminate the contingencies that limit effective prevention and health promotion (Willms *et al.* 1992).

These points are illustrated by a study using face to face interviews which replicated the questions used in the selected paper but also included more open questions asking for the *reasons* behind the answers given to the forced-choice questions (Stott and Pill, 1990). The opportunity to qualify such answers revealed that the general practitioner's interest in lifestyle was regarded as legitimate only within certain limits and the importance of the nature of the doctor–patient relationship emerged clearly. It was concluded that issues of lifestyle can seem appropriate if clearly related to the presenting problem but risk factor counselling which is unrelated and unrequested is less likely to be acceptable and indeed runs against the basic premise of patient-centred care. Doctors should not therefore assume they are welcome to give unsolicited advice. These conclusions may appear to contradict the requirements of the 1990 contract which lays obligations on doctors to provide preventive advice and achieve specified targets of population coverage. Stott (1993) argues that the new contract reflects the systematic attempt by government to force general practice to focus on the needs of the population rather than the individual and that there are inherent dangers for family medicine in this shift of policy. In his view 'If general medical practitioners let go of their responsibility and accountability to the individual they become authoritarian public health doctors and risk losing their credibility with their patients'. This is a debate which shows little sign of immediate resolution and I would confidently predict that we may expect future papers on how, why, whether, by whom, and even, if, health promotion should be carried out in the general practice consultation.

To conclude, the selected paper is a good example of the dominant research approach in general practice derived from the natural sciences, and can only properly be criticized for the extent to which it does or does not meet the criterion of reliability and validity that paradigm lays down. Different models of research not only use different methods but also frame different research questions and use different criteria for validity. The new series of monographs on research methods for primary care (Norton *et al.* 1991; Crabtree and Miller 1992; Tudiver *et al.* 1992) provides examples of an eclectic approach to meet the research and evaluation needs of the developing discipline of primary care. There is now an increasing lobby for the view that qualitative research has a particular relevance for family medicine/general practice (Kuzel 1986; Murphy and Mattson 1992); and the view is put forward here that there is certainly potential for a fresh approach to the issue of health promotion in general practice.

REFERENCES

Boulton, M. and Williams, A. (1983). Health education in the general practice consultation: doctors' advice on diet, alcohol and smoking. *Health Education Journal*, **42**, 57–63.

Calnan, M. and Johnson, B. (1983). Influencing health behaviour: how significant is the general practitioner? *Health Education Journal*, **42**, 34–5.

Coulter, A. and Schofield, T. (1991). Prevention in general practice: the views of doctors in the Oxford region. *British Journal of General Practice*, **302**, 1057–60.

Crabtree, B. F. and Miller, W. L. (1992). *Doing qualitative research*. Sage Publications, London.

Department of Health and Social Security (1976). *Prevention and health: everybody's business*. Her Majesty's Stationary Office, London.

Department of Health and Social Security (1977). *Prevention and health: reducing the risk: safer pregnancy and childbirth*. Her Majesty's Stationary Office, London.

Doll, R. and Peto, R. (1983). Prospects for prevention. *British Medical Journal*, **286**, 445–52.

Handel, S. (1973). Changing smoking habits in general practice. *Postgraduate Medical Journal*, **49**, 679–681.

Jamrozik, K. and Fowler, G. (1982). Anti-smoking education in Oxfordshire general practices. *Journal of Royal College of General Practitioners*, **32**, 179–83.

Jamrozik, K., Vessey, M., Fowler, G., Wald, N., Porter, G. and van-Vunakis, H. (1984). Controlled trial of three different anti-smoking interventions in general practice. *British Medical Journal*, **288**, 1499–503.

Kuzel, A. K. (1986). Naturalistic inquiry: an appropriate model for family medicine. *Family Medicine*, **18**, 369–74.

Murphy, E. and Mattson, B. (1992). Qualitative research and family practice: a marriage made in heaven? *Family Practice*, **9**, 85–91.

Norton, P. G., Stewart, M., Tudiver, F., Bass, M. J., and Dunn, E. V. (1991). *Primary care research: traditional and innovative approaches*. Sage Publications, London.

Patton, M. Q. (1987). *How to use qualitative methods in evaluation*. Sage Publications, London.

Pill, R. M. & Stott, N. C. H. (1987). The stereotype of 'working class fatalism' and the challenge for primary care health promotion. *Health Education Research*, **2**, 105–14.

Pill, R. M., Jones, E. G., and Stott, N. C. H. (1989). Opportunistic health promotion: quantity or quality? *Journal of the Royal College of General Practitioners*, **39**, 196–200.

Porter, A. M. W. and McCullough, D. M. (1972). Counselling against cigarette smoking: a controlled study from general practice. *Practitioner*, **209**, 686–90.

Royal College of General Practitioners (1981*a*). *Health and prevention in primary care*. RCGP Report from General Practice, London.

Royal College of General Practitioners (1981*b*). *Prevention of arterial disease in general practice*. RCGP Report from General Practice 19, London.

Royal College of General Practitioners (1981*c*). *Prevention of psychiatric disorders in general practice*. RCGP Report from General Practice 20, London.

Royal College of General Practitioners (1981*d*). *Family planning—an exercise in preventive medicine*. RCGP Report from General Practice 21, London.

Royal College of General Practitioners (1982). *Healthier children—thinking prevention*. RCGP Report from General Practice 22, London.

Royal College of General Practitioners (1983). *Promoting preventive medicine*, occasional Paper, 22. RCGP Report from General Practice, London.

Russell, M. A. N., Wilson, C., Taylor, C., and Baker, C. D. (1979). Effect of general practitioners' advice against smoking. *British Medical Journal*, **281**, 231-5.

Russell, M. A. N., Stapelton, J. A., Jackson, P. H., Hajek, P., and Belcher, M. (1987). District programme to reduce smoking: effect of clinic supported brief intervention by general practitioners. *British Medical Journal*, **295**, 1240-4.

Sanders, D., Fowler, G., Mant, D., Fuller, A., Jones, L., and Marzillier, J. (1989). Randomised controlled trial of anti-smoking advice by nurses in general practice. *Journal of Royal College of General Practitioners*, **39**, 273-276.

Silagy, C., Muir, J., Coulter, A., Yudkin, P., and Thorogood, M. (1992). Lifestyle advice in general practice: rates recalled by patients *British Medical Journal*, **305**, 871-4.

Smail, S. (1982). Opportunities for prevention: the consultation. *British Medical Journal*, **284**, 1092-3.

Stott, N. C. H. (1983). *Primary health care: bridging the gap between theory and practice*. Springer-Verlag, Berlin.

Stott, N. C. H. (1993). 'When something is good, more of the same is not always better'. *British Journal of General Practice*, **43**, 254-8.

Stott, N. C. H. and Davis, R. H. (1979). The exceptional potential in each primary care consultation. *Journal of the Royal College of General Practitioners*, **29**, 201-5.

Stott, N. C. H. and Pill, R. M. (1990*a*). 'Advise Yes, Dictate No? Patients' views on health promotion in the consultation. *Family Practice*, **7**, 125-31.

Stott, N. C. H. and Pill, R. M. (1990*b*). *Making changes: a study of working-class mothers and the changes made in their health-related behaviour over 5 years*. University of Wales College of Medicine, Cardiff.

Tuckett, D., Boulton, M., Olson, C., and Williams, A. (1985). *Meetings between experts: an approach to sharing ideas in medical consultations*. Tavistock, London.

Tudiver, F., Bass, M. J., Dunn, E. J., Norton, P.G., and Stewart, M. (1992). *Assessing interventions*. Sage Publications, London.

Wallace, P. G. and Haines, A. P. (1984). General practitioners and health promotion: what patients think. *British Medical Journal*, **289**, 534-536.

Wallace, P. G., Brennan, P., and Haines, A. P. (1987). Are general practitioners doing enough to promote healthy lifestyle: findings of the Medical Research

Council general practice research framework study on lifestyle and health. *British Medical Journal*, **294**, 940–2.

Wallace, P. G., Cutler, S., and Haines, A. (1988). Randomised controlled trial of general practitioner intervention in patients with excessive alcohol consumption. *British Medical Journal*, **297**, 663–8.

Williams, A. and Boulton, M. (1988). Thinking prevention: concepts and constructs in general practice. In: *Biomedicine examined* (ed. M. Lock and D. R. Gordon), pp. 227–55. Kluwer Academic Publishers.

Willms, D. G., Johnson, N. A., and White, N. A. (1992). A qualitative study of family practice physician health promotion activities. In *Doing qualitative research* (ed. B. F. Crabtree and W. L. Miller). Research methods for primary care Vol. 3. Sage Publications, London.

World Health Organisation (1978). *Declaration of Alma Ata*. Report on the International Conference on Primary Health Care, Alma Ata, USSR. Geneva.

Zyzanski, S. J., McWhinney, I. R., Blake, R., Crabtree, B. F., and Miller, W. L. (1992). Qualitative research: perspectives on the future. In *Doing qualitative research* (ed. B. F. Crabtree and W. L. Miller). Research methods for primary care Vol. 3. Sage Publications, London.

General practitioner and health promotion: what patients think

Paul G. Wallace, Andrew P. Haines
Dept. of General Practice, St. Mary's Hospital Medical School
BMJ, **289**, 534-6 (1984)

Abstract

Although there has been growing interest in general practitioners' participation in promoting health, little is known about the attitudes of their patients. Thus we sent a copy of a self administered questionnaire (the Health Survey Questionnaire) to 3452 patients aged 17-70 who were registered with two practices in north west London. Questions about attitudes to and perceptions of general practitioners' interest in weight, smoking, drinking, and fitness problems were included. The patients were also asked whether they thought that they had a problem in any of these areas.

The response rate was 72%. Of those who responded, the proportions who thought that their general practitioners should be interested ranged from 72% in the case of fitness to 83% for weight, but only 38% thought that general practitioners had in fact been interested in fitness and only 48% thought so about weight. Forty one per cent of the respondents thought that they had a fitness problem, 42% a weight problem, and 59% of the smokers thought that they had a smoking problem. Four per cent of respondents stated that they had a drinking problem. Of those patients who said they had a problem, the proportions who thought that their general practitioners had seemed interested ranged from 43% for fitness to 69% for smoking.

The findings of this study suggest that greater participation by general practitioners in health promotion would be well received by most patients and that currently there may be considerable discrepancies between patients' expectations and their perceptions of their general practitioner's interest in these areas of preventive medicine.

INTRODUCTION

There has been much interest in the potential for the participation of general practitioners in promoting a healthy lifestyle.[1] This has arisen both because of the accumulation of evidence of the harmful effects of cigarette smoking, over-weight, excessive alcohol consumption, and lack of physical exercise, and because of the recent findings that intervention by general practitioners can result in important changes in behaviour in the case of cigarette smoking.[2-6] Though the effectiveness of intervention for the other three areas has yet to be adequately evaluated, general practitioners are already coming under increasing pressure to incorporate health promotion of this kind into their routine clinical work.[7]

There is little recent information on what patients think about the practice of health promotion by their family doctors. We have therefore carried out a study to determine whether patients think that it is appropriate for general practitioners to be interested in problems relating to four areas of lifestyle – namely, smoking, weight, drinking, and fitness. In addition, we have examined the patients' perceptions both of their own health problems in these areas and of the interest actually shown by their general practitioners.

PATIENTS AND METHODS

The study population consisted of all the patients aged 17 to 70 years inclusive who were registered with two general practices in a health centre located on a council housing estate in Harlesden, London. Of the 3997 patients on the age-sex register of the two practices, 545 (14%) had changed address without notifying the practice or left the practice or died. Current addresses were available for a total of 3452 patients.

The self administered questionnaire which we designed for this study, the Health Survey Questionnaire, included questions relating to weight, cigarette smoking, alcohol consumption, and fitness. A simple multiple choice of four responses was used throughout. The three sections relevant to this paper are:

(*1*) *In your opinion should your family doctors be interested in*: (*a*) weight problems, (*b*) smoking problems, (*c*) drinking problems, (*d*) fitness problems?

(*2*) *From what you know of your family doctors, have they seemed interested in*: (*a*) weight problems, (*b*) smoking problems, (*c*) drinking problems, (*d*) fitness problems? The possible responses to these questions were: definitely, probably, no, don't know.

(*3*) *Do you think you have a*: (*a*) weight problem, (*b*) smoking problem, (*c*) drinking problem, (*d*) fitness problem? The possible responses to these questions were: definitely, possibly, probably not, definitely not.

A reply paid copy of the questionnaire was initially mailed to each patient in the study, accompanied by a covering letter signed by their general practitioners. Reminder letters were sent at three and six weeks, the latter including a fresh reply paid questionnaire. At two months the records of the persistent non-responders were tagged so that they could be identified on attending surgery and personally handed a copy of the questionnaire by one of their own general practitioners. Questionnaire responses were coded and double checked before computer analysis. χ^2 testing and Armitage χ^2 linear trend analysis were used for comparing results in different groups.

In addition to the patient questionnaire the five general practitioners who worked in the two practices each received a copy of a questionnaire that asked whether they had regularly given advice on smoking, weight, drinking, and fitness in their consultations over the 12 months before the start of the study. The responses to these questionnaires were analysed manually.

RESULTS

The general practitioner questionnaires were returned satisfactorily completed. All five general practitioners stated that they had regularly given advice on weight, smoking, drinking, and fitness as part of their routine work during the 12 months before the study started. Of the 3452 patients included in the study, 2477 (72%) returned a questionnaire within six months of the original mailing. Questionnaires were received from 1394 of the 1859 women (75%) and from 1069 of the 1593 men (67%). Ten of the returned questionnaires had been incompletely identified before mailing and four others were returned with no decipherable response so the analysis is based on the remaining 2463 questionnaires. The response rates for men and women were significantly different (p < 0.001) and there was a significant trend for a greater response rate with increasing age for both sexes (p < 0.01).

Responses to: Should your general practitioner be interested? (*Table I*)—The

Table I *Should your general practitioners be interested?*

	Weight problems		Smoking problems		Drinking problems		Fitness problems	
	No (%) of men	No (%) of women	No (%) of men	No (%) of women	No (%) of men	No (%) of women	No (%) of men	No (%) of women
Should be interested	793† (81)	1098† (85)	777 (79)	1009 (81)	744* (77)	978* (81)	689 (72)	888 (73)
Should not be interested	128 (13)	97 (8)	149 (15)	143 (11)	163 (17)	153 (13)	179 (19)	200 (16)
Don't know	63 (6)	98 (7)	57 (6)	92 (8)	54 (6)	77 (6)	91 (9)	136 (11)

Comparison of men with women: *p < 0.05; †p < 0.001.

Table II *Have your general practitioners seemed interested?*

	Weight problems		Smoking problems		Drinking problems		Fitness problems	
	No (%) of men	No (%) of women	No (%) of men	No (%) of women	No (%) of men	No (%) of women	No (%) of men	No (%) of women
Have seemed interested	461 (47)	574 (48)	528† (54)	595† (48)	425† (44)	449† (38)	392* (41)	437* (36)
Have not seemed interested	193 (20)	224 (19)	158 (16)	199 (16)	198 (21)	233 (19)	209 (22)	249 (20)
Don't know	325 (33)	393 (33)	290 (30)	444 (36)	335 (35)	515 (43)	358 (37)	545 (44)

Comparison of men with women: *p < 0.05; †p < 0.01.

completion rates for the questions in this section ranged from 88% to 93%. The overall proportions of respondents who thought that their general practitioners should be definitely or probably interested ranged from 72% for fitness problems to 83% for weight problems. The proportions for smoking and drinking problems were 81% and 80% respectively. Significantly more women than men thought that their general practitioners should be interested in weight and drinking problems, but there was no significant sex difference between the responses for smoking and fitness. There were significant age trends for both sexes in the case of attitudes towards general practitioner interest in weight problems, with larger proportions of younger patients in favour. For men, the proportions decreased progressively from 79% in the age group 21 to 30 to 65% in the age group 61 to 70 (p < 0.05), while for women the proportions decreased from 84% in the age group 17 to 20 to 67% in the age group 61 to 70 (p < 0.001). For drinking and fitness problems there were similar age trends for women only (p < 0.05) in both cases. There was no significant trend for either sex for smoking.

Responses to: Have your general practitioners seemed interested? (Table II) — The completion rates for the questions in this section ranged from 87% to 90%. The proportions of respondents who thought that their general practitioners had definitely or probably seemed interested ranged from 38% for fitness problems to 51% for smoking problems. The proportions for weight and drinking problems were 48% and 41% respectively. Significantly more men than women thought that their general practitioners had seemed interested except for weight problems. The proportions of respondents who were unsure about their general practitioners' interest ranged from 33% for weight problems to 41% for fitness problems, while for those who thought that their general practitioners had not seemed interested the range was from 16% in the case of smoking problems to 21% for fitness problems.

Responses to: Do you think you have a problem? (Figure) — Completion rates for the questions in this section ranged from 86% to 92%. Forty two per cent of respondents thought that they had a weight problem and 41% a fitness problem. Twenty four per cent stated that they had a smoking problem (43% of the respondents smoked and 59% of these thought that they had a problem). Four per cent of patients thought that they had a drinking problem (67% of the respondents stated that they had taken alcohol in the three months before receiving the questionnaire, and 6% of the drinkers felt they had a problem). Significantly more women than men had weight and fitness problems by their own assessment, and significantly more men than women had a drinking problem. The only significant age trend was for men with a weight problem, the proportions increasing progressively from 14% in the age group 17 to 20 to 34% in the age group 51 to 60 (p < 0.005).

Patients with and without a problem by their own assessment: attitudes to and perceptions of general practitioner interest — Of the patients who stated that they had a problem, the proportions who thought that their general practitioners should be interested ranged from 79% for drinking problems to 91% for weight problems, with 89% for smoking and 3% for fitness problems. Except for drinking, these proportions were significantly higher (p < 0.005) than for the patients who did not feel that they had a problem. The proportions of patients with a problem who thought that their general practitioners had seemed interested

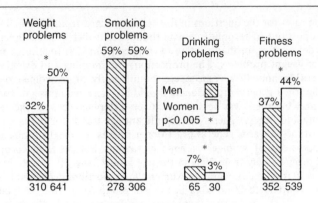

Patients who thought that they had a problem (definitely/possibly). Numbers of patients with smoking problems are expressed as proportions of smokers.

ranged from only 43% for fitness to 69% for smoking, with 62% for weight and 63% for drinking. These proportions were all significantly higher than for the patients who did not think that they had a problem ($p < 0.05$). None the less, between 11% and 17% of the patients with a problem thought that their general practitioners had not seemed interested and a further 20% to 40% were unsure.

DISCUSSION

The overall response rate of 72% to the questionnaire compares favourably with similar studies where rates have ranged from 40% to 59%.[8-10] Together with the generally high completion rates for individual questions, this indicates that our Health Survey Questionnaire was well accepted by patients in general practice. The higher response rates among women and older patients may be the result of their greater interest in health and related topics. These patients, however, also tend to have higher consultation rates and be better acquainted with their general practitioners and thus may be more inclined to participate in such studies.

Most of the patients expected their general practitioners to be interested in weight, smoking, drinking, and fitness problems, with small though statistically significant differences between the proportions for the four areas. Similar results were found in a UK study undertaken some time ago where 80% of patients in general practice expected to be given helpful advice on health at the general practitioner's surgery or the clinic, and in the United States in a more recent study where 90% of family practice patients expected to receive encouragement from their family doctor to lose weight, give up smoking, and take more exercise.[8,9] The observation that younger patients, especially women, have a greater expectation of general practitioner interest in weight, fitness, and drinking problems indicates that a change in attitude may have occurred with time. Because of the cross sectional nature of the study, however, it is not possible to distinguish between a cohort effect, with those born in more recent years having a greater expectation of general practitioner interest and a decline in patients' expectation of general practitioner interest with increasing age. In view of the lower response rates among younger age groups it is also possible that the younger patients with lower expectations of general practitioner interest were underrepresented. Whatever the explanation, the absence of an age trend for smoking indicates a longstanding and widespread awareness of the association between smoking and ill health.

Substantial proportions of patients felt that they had a weight, smoking, or fitness problem, and 4% thought that they had a drinking problem. Despite the statements by all five general practitioners that they had regularly given advice in these four areas, between 31% and 57% of these patients either thought that their general practitioners were not interested or were unsure about their interest. Although further information is required about the validity of the patients' concern about their health in each of the four areas, these findings suggest that there may be considerable scope for increasing the emphasis on health promotion in general practice. The finding that higher proportions of men thought that their general practitioners had been interested may reflect not only the larger proportions of men with objective evidence of cause for concern, but also the awareness of their general practitioners of the increased risks of ischaemic heart disease in men compared with women.

The effectiveness of general practitioner intervention has only been properly evaluated for smoking. Since the evidence from our study suggests that there is widespread demand among patients for general practitioners to become concerned in all four areas of health, and there is evidence that such expectations are largely shared by the general practitioners themselves, carefully designed trials of intervention in the three other areas should now be carried out.[10]

ACKNOWLEDGEMENTS

We thank the Sir Jules Thorn Charitable Trust and the Royal College of General Practitioners for the award of the fellowship in general practice, our colleagues Drs Bowman, Gellert, and Teuten for their kind cooperation, and Angela Booroff, Eva Goldenberg, and Janet Stride for their help with the administration of the study. We also thank Richard Allen, Patrick Brennan, and Douglas Jones for help with the data analysis and the Brewers' Society for financial support.

REFERENCES

1. *Health and Prevention in Primary Care*. Report from General Practice No 18. London: Royal College of General Practitioners, 1981.
2. Doll, R., Peto, R. Mortality in relation to smoking: 20 years' observation of male British doctors. *Br Med J* 1976, ii, 1525-36.
3. Rhoads, G. G., Kagan, A. The relation of coronary disease, stroke and mortality to weight in youth and in middle age. *Lancet* 1983, i, 492-5.
4. Marmot, M. G., Rose, G., Shipley, M. J., Thomas B. J. Alcohol and mortality—a U-shaped curve. *Lancet* 1981, i, 580-3.
5. Kannel, W. B., Sorlie, P. Some health benefits of physical activity. The Framingham study. *Arch Intern Med* 1979, **139**, 857-61.

6. Russell, M. A. H., Wilson, C., Taylor, C., Baker, C. D. Effect of general practitioners' advice against smoking. *Br Med J* 1979, **ii**, 231-5.
7. Skinner, H. A., Holt, S. Early intervention for alcohol problems. *J R Coll Gen Pract* 1983, **33**, 787-92.
8. Pike, L. A. The consumer demand for health education. *J R Coll Gen Pract* 1971, **21**: 156-60.
9. Hyatt, J. D. Perceptions of the family physician by patients and family physicians, *J Fam Pract* 1980, **ii**, 295-300.
10. Catford, J. C., Nutbeam, D. Prevention in practice: what Wessex general practitioners are doing. *B Med J* 1984, **288**, 832-4.

5 Description and experiment in a single practice

John Bain

INTRODUCTION

In an account of his observations on general practice in the United Kingdom, Donabedian (1986) remarked on the undercurrent of uncertainty about what general practice really is. Despite the growth of the Royal College of General Practitioners, the creation of university departments of general practice, the vocational training system, and a widespread desire to to count the countable, much of what occurs in the consulting room remains resistant to easy measurement. At a time of great change in the Health Service (Secretaries of State for Health 1989), where the main focus often seems to be on matters of organization and finance, audit, and accountability, there is a place for a reminder of the true content of the consultation in general practice.

In his work on the 'Temporarily dependent patient' (Thomas 1974) and the 'Therapeutic illusion' (Thomas 1978), Thomas set out to explore areas of care which were seldom given prominence in the literature of general practice. His study spanned a period of time when successive editions of College publications on *Trends in general practice* (RCGP 1977, 1979) were being produced. While these important publications provided excellent background information on workload, prescribing, premises and organization, the health team, and education for general practice, few of the contributors attempted to describe the large areas of daily practice which create the greatest uncertainty for the general practitioner.

I have to confess to an attraction to the ideas that Thomas has developed in his papers (1974, 1978), where he concentrated on the plethora of undifferentiated problems which are faced in daily practice. The traditional medical model of education and training emphasizes the need for precise pathological diagnoses which, in theory, will lead to well delivered treatments, resulting in cure. This is very important in large areas of medicine, where the health professional is dealing primarily with the underlying disease rather than in the experience of illness. It is of greatest importance in medicine practised in hospital and has less application in primary care.

New entrants to general practice are often overwhelmed by the variety of ills that are shared in the consulting room; Thomas takes us into this territory, where his underlying philosophy is based on the belief that for many patients a 'wait and see' policy will result in no harm to those who are seeking advice from their doctor. He sets out to ascertain the proportion of people who do not fit into neat diagnostic categories and to test the theory that the prescription of medicines may be counter-productive. How well does he achieve these aims?

1. TEMPORARILY DEPENDENT PATIENT IN GENERAL PRACTICE

The starting point for this descriptive study was Thomas' own practice in Waterlooville on the northern outskirts of Portsmouth, where he was in a four partner practice, looking after around 10 000 patients. In this study he classified 5409 consultations, undertaken over a one year period, into three broad categories:

1. Service patients (1561 = 29 per cent) who received services such as immunization, contraceptive services, and routine medical checks;
2. Diagnoses (2192 = 40 per cent). Patients with specific symptoms and signs which resulted in a definite diagnosis';
3. Undiagnosed patients (1656 = 31 per cent) or 43 per cent of those presenting with illness which he described as those with ill-defined problems for which no traditional diagnosis could be made.

Of the 1656 patients who fell into this latter category of ill-defined problems, and who received no definite physical treatment 1191 (72 per cent) had not returned to see him within one month of the original consultation and were assumed 'to have recovered'. This group of patients was followed up at three months and two years after the initial consultation, when a 10 per cent sample completed a postal questionnaire on the outcome of their original complaint. No details are given about the age distribution of the patients studied.

These patients with undifferentiated problems were asked two specific questions:

1. Did you get better?
2. Did you have any further treatment?

Data are then presented on a total of 1107 patients, of whom 976 stated that they had got better, and 131 said that they had not sought further treatment, even though they had not got better. To ensure that the findings for the undiagnosed group were not peculiar to that group for other reasons, they were compared with the patients in whom a diagnosis had been made, by selecting 500 patients from each and comparing responses to the Eysenck personality inventory and the Millhill vocabulary scale. In addition, the two groups were compared for age, sex, social class, length of registration with the practice, and the number of consultations and absence from work in the four years prior to the investigation.

Thomas found that there was no significant difference between the patients with ill-defined problems and those for whom a definite diagnosis had been made, in regard to all the factors assessed on the psychological inventories, including neuroticism, introversion, and intelligence.

There are, however, questions about the response rate both in the groups followed up after the initial consultation and in the sample of diagnosed and undiagnosed patients, to whom the personality and vocabulary scales were applied. Low response rates may well be a source of bias, in which case the characteristics of non-responders as well as responders to questionnaires of this kind need to be established.

There also have to be questions about the validity of categorizing patients into 'ill-defined' and 'diagnosed' groups by a single general practitioner, particularly the same one who has posed the original research question. A more rigorous approach in a study of this kind might be to use a panel of colleagues to review case records in order to validate the researcher's classification of patients into one or other group. None the less, despite these difficulties, Thomas' study highlights the fact that a substantial proportion of work in general practice is concerned with problems that cannot readily be classified in the traditional manner, nor adequately explained by the conventional biomedical model.

His conclusions are acceptable when considering individual consultations, but general practice often involves repeated contacts with patients over a long period of time. There is little clarity in this paper about the number of patients who already had established diagnoses and were merely consulting with symptoms and signs superimposed on an underlying condition to which a disease label had already been given. The simplicity of his follow-up questionnaire, simply asking whether patients had got better and whether they received any further treatment, does not permit a more detailed analysis of the characteristics of their individual problems, in order to determine the answer to this question, nor does it explore the interesting, allied issue of the use of lay networks, including family and friends, and other providers of primary care, to resolve their problems.

2. THE CONSULTATION AND THE THERAPEUTIC ILLUSION

Thomas' interest in undifferentiated problems in general practice extended to an investigation of 200 undiagnosed patients who were randomly allocated to one or two treatments, using a randomized controlled trial design. The two groups were as follows:

1. Patients who were told that no illness had been found and that they required no treatment.

2. Patients who were given a symptomatic diagnosis and medication.

The outcome in both groups was assessed by measuring the numbers who did not return within one month, or returned with the same complaint within one month, or who returned with a different complaint within one month. An additional assessment included a postal questionnaire, enquiring whether patients had got better and whether they had required further treatment. The two groups were similar in terms of age and other demographic features and Thomas was not able to find any significant differences in terms of the outcomes which he measured. From these results he argued that there was a strong case for giving no treatment (i.e. a prescription) to people with non-specific symptoms. It could, of course, also be argued that giving treatment was equally effective, since 30 per cent of the patients in both the treated and non- treated groups returned within a month with a similar or different complaint. However, these data, highlight the dilemmas facing the practising doctor, dilemmas which are underlined by Thomas' other finding that, in both groups, 30 per cent had not recovered and had received further treatment.

The interpretation of this work presents a number of problems. As in all single practice, single doctor studies, personal involvement in conceiving, designing, and carrying out the study has to be considered as a potential source of bias. However, the results presented in this paper fit in with my own belief systems, and I am able to resonate with Thomas's plea that 'modern general practice suffers from an unjustified enthusiasm for treatment on the part of both patients and doctors, which results in crowded surgeries, too little time for the ill patient, a large and ever-growing drug bill, increasing iatrogenic disease, and the proliferation of illness' (Thomas 1978). These are words that are well worth emphasizing fifteen years on when many general practitioners are sceptical of a new contract which may lead to primary care attempting to mimic the medical model of hospital practice.

Further difficulties in interpreting this work relate to the extent to which single-practice studies of this kind can be more widely extrapolated and generalized. Thomas has used a simple measure of outcome, which may be too simplistic, and the patients who were studied have already become used to his style of practice, and may well have hidden complaints that would not be explored by his 'hands off' approach. The extent to which symptoms such as abdominal pain, cough, headache, chest pain, and dizziness were the early signs of disease is difficult to unravel, and it is worth noting that a third of patients in the second study, irrespective of management, still had symptoms one month after the initial consultation. The reader is given no indication about the eventual outcome for these patients; a longer-term follow-up might well have revealed clinical developments which might have necessitated a re-evaluation of some of the conclusions drawn from the studies.

I would also take issue with his statement about 'no treatment'; in his introductory sections to both papers, he refers to the therapeutic effect of the 'drug doctor' (Balint 1964), and there seems little doubt that Thomas' own approach in terms of explanation and reassurance will have had a powerful influence on his patients. McWhinney (1989) suggested that the essential basis for effective reassurance is a trusting relationship between patient and doctor, and Thomas seems to have had this in abundance. However, I would accept Thomas' emphasis on the need to recognize that up to four out of ten patients have clusters of symptoms to which traditional diagnostic labels cannot readily be given. To explore his contention that less invasive intervention may be as effective as prescribing would probably require a much longer period of follow-up, accompanied by research methods which investigated more rigorously the patient's interpretation of health care outcomes, as well as a more detailed and thorough clinical follow-up. It may, of course, be that an extended study of this kind would not lead to significantly different results.

THE RELEVANCE OF THE STUDIES

Despite some reservations about methodology and presentation of data, Thomas' work serves as a perennial reminder of how a good question can be posed and explored within the setting of a single general practice. Howie (1989) considers this the first step in research, and one so important that it merited a full chapter in his book on research in general practice. Thomas' questions satisfy Howie's criteria in that he has sought to find answers about an important feature of a general practitioner's work in the content of the consultation and has pursued an idea to which reasonable answers can be obtained within a relatively short period of time. Thomas was a general practitioner working on his own with no research assistants and no research grants, producing important information about the content of general practice. His work would be on my required reading list for all students and trainees. He reminds us that human beings are sturdy, and evidence that encourages us to adopt a 'wait and see policy' is essential when medical practice is often dominated by the pressure to prescribe. His research methods may have limitations but he highlights the fact that medicine can be imprisoned in the belief that a diagnosis is central to decision making.

The current literature in general practice is full of ideas on extending the primary care team, developing business plans, and creating clinics, in the belief that this will improve outcomes for patients. Thomas' argument that minimal interference will not result in serious omissions in care tends to go against the current trend to search more actively for 'hidden disease'. I suspect that if his conclusions are to be accepted, then more attempts to bring conformity to diagnostic and therapeutic activity in general practice are doomed to failure.

In a more recent publication, Thomas reminds us that 'Examination of general practice at any time in the last thousand years shows doctors holding firm convictions about diagnosis and treatment which were later shown to be no more than interpretations of illness made in the light of the beliefs of the times (Thomas 1981). The curiosity he has displayed is a stimulus to escape from the straightjacket of 'The therapeutic illusion' and his work captures the flavour as well as the substance of day-to-day work in general practice. It would be interesting to discover if the content of general practice in the 1990s has changed substantially since the 1970s. I suspect not, and that temporarily dependent patients are still alive and well throughout general practice.

REFERENCES

Balint, M. (1964). *The doctor, his patient and the illness*. Pitman Medical.

Donabedian, A. (1986). *In pursuit of quality*. Chapter 5. Royal College of General Practitioners.

Howie, J. G. R. (1989). *Research in general practice*. 2nd edn. Chapman and Hall.

McWhinney, I. R. (1989). *Textbook of family medicine*. Oxford University Press.

Royal College of General Practitioners (1977, 1979). *Trends in general practice*. British Medical Journal.

Secretaries of state for Health, Wales, Northern Ireland and Scotland (1989). *Working for patients*. (CM555), HMSO, London.

Thomas, K. B. (1974). Temporarily dependent patient in general practice. *British Medical Journal*, **1**, 625-6.

Thomas, K. B. (1978). The consultation and the therapeutic illusion. *British Medical Journal*, **1**, 1327-8.

Thomas, K. B. (1981). Attitudes to psychological illness in general practice. *British Medical Journal*, **282**, 1157-8.

Temporarily dependent patient in general practice

K. B. Thomas
GP, Portsmouth

BMJ, **1**, 625-6 (1974)

Abstract

Of 3,848 consultations with patients to 330 general practice surgeries during one year, no diagnosis was made in 1656. The latter received no effective treatment other than contact with their doctor, and were asked to return if they did not feel better. But 1191 did not return. Subsequent inquiry showed that 976 (82%) said they had been made better, and a further 131 (11%) said that, though they were no better, they had not sought further treatment.

The 'successfully untreated' patients were shown not to differ significantly from those patients in whom a definite diagnosis had been made, with regard to neuroticism, extraversion, intelligence, age, sex, marital status, social class, length of stay in the practice, number of consultations, and absence from work. These patients have been called 'temporarily dependent' patients and their possible influence on diagnosis is discussed.

INTRODUCTION

Practitioners view psychological illness in many different ways.[1-9] An examination of 25 surveys in general practice showed that the reported incidence of psychiatric consultations varies from 3% to 65%, and that the rates are lowest (3%-12%) in surveys by doctors investigating general morbidity, and highest (24%-65%) in surveys by doctors assessing only psychological illness.[10]

General morbidity surveys have shown that there is also a variation in the incidence of many physical illnesses. Shepherd showed that in 14 practices there was a significant variation in the reported rate for nearly every category of illness, while in respiratory illness in female patients the variation exceeded that for psychiatric disorders.[3] A similar variation has been found in other surveys.[8, 11-13] Thus the validity of much diagnosis in general practice seems to be in doubt.

This paper attempts to show two factors which are thought to make accurate diagnosis difficult. The first is the existence of those patients who do not seem to have evidence of illness, and the second is that many of them get better without any treatment other than contact with their doctor, and this encourages him to believe that his treatment has been effective and his original diagnosis correct.

PATIENTS AND METHODS

The investigation was made by one partner in a group practice of four, covering 10 300 patients, in a suburban area of Hampshire (tables I-II). The patients were divided into the following groups: firstly, 'service' patients—those who came for services; they were not ill, and came for inoculations, cervical smears, the pill; secondly, the 'diagnosed' group—patients who presented with definite signs and symptoms of physical or psychological illness, and for whom a diagnosis was made, and thirdly, the 'undiagnosed' group—patients who presented with little or no signs of physical or psychological illness, and for whom no definite

Table I Social Class of Waterlooville Electoral Ward compared with a Sample of 500 Patients who attended the Surgery during the Investigation

Social Class	Waterlooville Electoral Ward*	Surgery
	%	%
I	300 (7.7)	10 (2.0)
II	690 (17.7)	85 (17.0)
III	1,500 (38.8)	275 (55.0)
IV	470 (12.1)	78 (15.6)
V	190 (4.9)	6 (1.2)
Not classified	710 (18.3)	46 (9.6)

*1966 Census.

Table II Age of 1000 of the Practice Population, compared with the Waterlooville Electoral Ward (Census 1966) in Parentheses

Age:	0–14	15–44	45–64	65+
500 Females	137 (168)	223 (199)	95 (66)	45 (54)
500 Males	170 (152)	210 (192)	74 (92)	46 (65)
Total	307 (320)	433 (391)	169 (158)	91 (119)

diagnosis could be made. This investigation is concerned with the last group.

Though the 'undiagnosed' patients were seen during ordinary surgeries in a busy general practice, each consultation was standardized as far as possible and consisted of: history taking, when the patient was given time to describe his complaint; a physical examination; reassurance that there was no serious illness and that the patient would soon be well; and a request to return in one week if the patient did not feel better. These patients received no effective physical treatment. A few received nothing at all, and most received a placebo.

The record cards of all the untreated patients were examined and, if the patient had not returned during the month after the consultation, he was assumed to have got better ('successfully untreated' patient). If he had been seen by a doctor for any reason, either with the original complaint or with a different one, he was considered to have failed to get better ('failed' patient). The assumption that the 'successfully untreated' patients had got better was tested by making two separate surveys, three months and two years after the consultation. Patients in a 10% sample were asked by post whether or not they had got better and whether they had sought any further treatment.

The 'successfully untreated' patients were investigated by comparing them with those patients for whom a definite diagnosis had been made (the 'diagnosed' group). The Eysenck Personality Inventory Form B and the Mill Hill Vocabulary Scale Form I were used to estimate the level of neuroticism, introversion, extraversion, and intelligence of a sample of 500 patients from the combined groups. The two groups were also compared for age, sex, civil state, length of stay in the practice, number of consultations, and absence from work during the four years previous to the investigation.

RESULTS

During the course of the investigation 5409 consultations were made; 1561 not for illness, but for services, leaving 3848 consultations for supposed illness. Of these consultations, a definite diagnosis was made in 2192 (56.9%) (the 'diagnosed' group), and no diagnosis was made in the remaining 1656 (43%) (the 'undiagnosed' group).

Of the patients in the 'undiagnosed' group 1191 (71.9%) had not returned to their doctor. The remainder, who had returned to their doctor within one month, were considered to have 'failed' in various degrees (table III).

Table III *Results of non-treatment*

	Numbers	Undiagnosed group %	Consultation for treatment %	Total consultations %
Patients who did not return . .	1,191	71.9	30.9	22
Patients who returned with the same complaint	228	13.8	5.9	4.2
Patients who returned with a different complaint	163	9.8	4.2	3.01
Patients who were untreated twice	64	3.8	1.6	1.1
Patients who returned for certificates	10	0.6	0.25	0.18

The sample of the untreated patients who did not return to the doctor were asked: 'Did you got better?' and 'Did you have any further treatment?' The results of both surveys are similar: 976 (82%) said they were better, and 131 (11%) said they had not sought further treatment even though they had not got better. This suggests that their concern for their symptoms may not have been very great, and indeed might have been further eased by the visit to the doctor.

No significant difference was found between the 'successfully untreated' patients, and those for whom a definite diagnosis had been made (the 'diagnosed' group), in regard to all the factors assessed such as neuroticism, introversion, intelligence, etc.

DISCUSSION

Most of this group of patients who received no treatment other than contact with their doctor improved. Moreover, they were asked to return to their doctor if they were no better, yet over 1000 did not. Most of them are still in the practice and, when asked, maintain that they got better and that they had no further treatment.

This investigation has shown that about two out of five patients coming for treatment show no objective evidence of physical or psychological illness. This is not a new finding,[14 15] and for instance the College of General Practitioners Research Committee, in a survey conducted in 11 practices, found that the average for 'firm diagnosis' was 55.5% with a range of 25.6%-72.4%.[13] Nevertheless, many general morbidity surveys in general practice show that a firm diagnosis is made for most patients.[16]

My 'successfully untreated' patients did not differ from patients attending the surgery with definite illness ('diagnosed group'). They were not a homogenous group with special characteristics, nor was there any objective evidence to show that they were suffering from any physical or psychological illness: they were just patients who, in their response—reasonable or unreasonable—to the ordinary ups-and-downs of life, had gone through a phase of temporary dependence, and could therefore be called 'temporarily dependent patients'.

Hence in much of medicine in general practice the doctor cannot make an accurate diagnosis, and patients are often made to feel better with no treatment other than contact with him. This is simple, but any attempt at a diagnosis in these patients is not a process of logical deduction from definite evidence, but rather his personal interpretation of an ill-defined and unstable situation.[17] So these patients may be viewed as suffering from organic disease, psychological illness, social stress, or more recently, from behavioural problems. But there is no need for this complexity; these patients are not ill in the accepted sense of the word, they are temporarily dependent and want only reassurance and support from their doctor.

ACKNOWLEDGEMENTS

I would like to thank Mr. J. R. Compton, Professor I. A. Forbes, and Dr. Ian Skottowe for their advice in writing this paper, and Dr. Stephen Mackeith for his encouragement and interest. I am grateful to Mr. D. J. Mulhall who was responsible for the statistical work and the computer programme.

REFERENCES

1. Dohrenwend, B. P., and Crandell, D. L., *American Journal of Psychiatry*, 1969-1970, **126**, 1611.
2. Marinker, M., *Journal of the Royal College of General Practitioners*, 1967, **XIV**, 59.
3. Shepherd, M., Cooper, B., Brown, A. C., and Kalton, G. G. W., *Psychiatric Illness in General Practice*, London, Oxford University Press, 1966.
4. Cooper, B., *Journal of Psychosomatic Research*, 1964, **8**, 9.
5. Kessel, N., and Shepherd, M., *Journal of Mental Science*, 1962, **108**, 159.
6. Lin, T., and Standley, C. C., *The Scope of Epidemiology in Psychiatry*, Geneva, World Health Organization, 1962.
7. Ryle, A., *Journal of the College of General Practitioners*, 1960, **3**, 313.
8. Howard, C. R. G., *Journal of the College of General Practitioners*, 1959, **2**, 119.

9. Taylor, Lord, and Chave, S., *Mental Health and Environment*, London, Longmans, 1964.
10. Thomas, K. B., MD Thesis, University of Liverpool, 1972.
11. Ward, T., Knowelden, J., and Sharrard, W. J. W., *Journal of the Royal College of General Practitioners*, 1968, **15**, 128.
12. Lees, D. S., and Cooper, M. H., *Journal of the College of General Practitioners*, 1963, **6**, 408.
13. Research Committee, *Journal of the College of General Practitioners*, 1958, **1**, 107.
14. Crombie, D. L., *Journal of the College of General Practitioners*, 1963, **6**, 579.
15. Eimerl, T. S., MD Thesis, University of Liverpool, 1962.
16. Williams, W. O., *Reports from General Practice XII*, London, Royal College of General Practitioners, 1970.
17. Shepherd, M., Cooper, B., Brown, A. C., and Kalton, G. W., *British Medical Journal*, 1964, **2**, 1359.

Contemporary themes: The consultation and the therapeutic illusion

K. B. Thomas
GP, Portsmouth

BMJ, 1, 1327–8 (1978)

Abstract
At 45 general-practice surgery sessions 200 patients in whom no definite diagnosis could be made were randomly selected for one of two procedures. Either they were given a symptomatic diagnosis and medication, or they were told that they had no evidence of disease and therefore they required no treatment. No difference in outcome was found between these two methods as judged by the return or not of the patient within one month and his statement that he did or did not get better.

INTRODUCTION

Diagnosis in general practice is frequently in doubt.[1-8] A firm diagnosis is made in only about half of all cases.[7-9] This article is concerned with those patients in whom no definite diagnosis can be made, who are not a homogeneous group.[10]

This inquiry investigated the consultation by omitting two of its traditional elements. A group of patients in whom no definite diagnosis could be made was told that they showed no evidence of disease and therefore required no treatment. The results of this procedure were compared with those of giving to a similar group of patients a symptomatic diagnosis and treatment.

PATIENTS AND METHODS

The investigation was made by one partner in a group of four covering 11 250 patients in a suburban area of Hampshire during 45 surgeries in September and October 1976. Patients attending general practice surgeries were divided into three groups: 'service' patients who were not ill and who came for services, such as innoculations, certificates, and cervical smears, the 'diagnosed' group, who presented with definite evidence of disease and in whom a firm diagnosis was frequently made; and the 'undiagnosed' group, who came with symptoms which were unsupported by definite evidence of physical or psychological illness and from which no definite diagnosis could be made.

A total of 200 undiagnosed patients attending during the investigation were randomly selected for one of two treatments: either they were told that as no evidence of illness had been found they required no treatment (the 'no treatment' group); or they were given a syptomatic diagnosis, often the one suggested by the patient, together with medication (the 'treatment' group). All the patients were asked to return in one week if they were no better. All the consultations by the undiagnosed patients in this inquiry were first consultations for each particular complaint. None of the undiagnosed patients was certified as unfit for work.

The outcome of the treatment and no treatment groups was assessed in two ways. Firstly, by noting whether the patient returned or not to one of the doctors within a month of the consultation, and if he did whether he had come with his same complaint or a different one. Secondly, by asking the patient one month after the consultation whether he had or had not got better, and whether he had required any further treatment. This second assessment was done by a postal survey and replies were eventually received from all.

RESULTS

During the investigation 635 consultation were made: 165 (26%) were not for illness but for services, leaving 470 consultations for supposed illness. Of these consultations, 270 patients (57%) showed definite evidence of disease (the diagnosed group) and no diagnosis was made in the remaining 200 (43%) (the undiagnosed group).

The outcome of treatment and no treatment of the undiagnosed patients, with regard to whether they did not return to a doctor within the month, or returned with the same or with a different condition, did not differ (table I).

The results of asking the patients whether they had got better and whether they required any further treatment showed that the two groups did not differ significantly (table II). There were not significant differences in the age distribution of patients in the various groups (table III). Nor was there any significant difference between the successfully and unsuccessfully untreated patients with regard to the frequency with which they consulted the author rather than his partners

Table I *Patients who did and did not return within a month to the doctor after treatment and no treatment*

	Treated	Untreated
Patients who did not return	72	71
Patients who returned with the same complaint	18	15
Patients who returned with a different complaint	10	14
Total	100	100

$\chi^2 = 0.95$;   DF = 2;   not significant

Table II *Results of asking patients whether they got better after treatment and no treatment*

	Treated	Untreated
Patients who got better	55	61
Patients who did not get better and had further treatment	30	29
Patients who did not get better and had no further treatment	15	10
Total	100	100

χ^2 square = 1.33;   DF = 2;   not significant

Table III *Age range in diagnosed, undiagnosed treated, and undiagnosed untreated patients*

	Age (years)			
	0–14	15–44	45–64	65–
Diagnosed	27	42	24	7
Undiagnosed treated	22	46	21	11
Undiagnosed untreated	29	44	23	4

$\chi^2 = 4.75$;   DF = 6;   not significant

in the five (and ten) consultations previous to the investigation or to their length of stay in the practice. Again, there was no significant difference between treated and untreated patients with regard to the number who consulted the author rather than his partners at the first consultation after the investigation.

Of 50 different symptoms complained of by the undiagnosed patients, the 10 commonest were abdominal pain (31); cough, (27); sore throat (20); headaches (11); pain in arm (10); pain in back (10); ears symptoms (9); pain in chest (9); pain in leg (S); and dizziness (7).

DISCUSSION

The doctor himself is a powerful therapeutic agent. In ancient times he was almost the only effective treatment and more recently Balint has described him as the most frequently used drug in general practice.[11]

The results of this study support the belief that the patient who is made better with no treatment will also be made better with treatment. The danger is that the doctor may ascribe recovery to his treatment and go on to see this as confirmation of his diagnosis. There may thus appear to be a relationship between diagnosis, treatment, and recovery which is not true. In the past this therapeutic illusion has been responsible for many mistaken diagnoses and much useless medication, and is probably responsible for a lot of unnecessary treatment today.

My inquiry has shown that in this practice two out of five patients coming for treatment do at least as well when they are told that they have no sign of disease and require no treatment as when they are given a disease label and conventional treatment. Giving a diagnosis and treatment where neither is indicated may encourage invalidism, and in whatever way patients are treated a model is made for their future behaviour in similar circumstances.

At the start of this investigation I was concerned about the reaction of patients to no treatment. These fears proved groundless. No one outwardly objected; a few patients expressed surprise; and most seemed to accept that if no disease had been found it was reasonable to give no treatment. Three mothers expresses satisfaction that their child was not going to receive another antibiotic. I also thought that no treatment would be more effective in those patients who had chosen to see me often in the past, or who had had a long stay in the practice; and that patients having received no treatment at one consultation would choose to see another doctor at the next consultation. Both these assumptions were proved wrong.

It would seem that patients tolerate no treatment better than doctors think they will. Surveys[12-14] have shown that 43–52% of patients expect to be given a prescription at a consultation, whereas most doctors think that the figures would be 80% or more.[15]

Modern general practice suffers from an unjustified enthusiasm for treatment on the part of both patients and doctors—which results in crowded surgeries, too little time for the ill patient, a large and ever-growing drug bill, increasing iatrogenic disease, and the proliferation of illness. Is it not strange that this enthusiasm for treatment in general practice is not matched by a corresponding interest in its outcome? Most patients get better and the therapeutic illusion tends to make doctors believe that their own particular method of treatment is responsible. Only when the results of different methods are compared with one another and with no treatment, will a scientific assessment of the consultation be possible.

The results of this inquiry suggests that a solution to the problem may be a change in the attitudes of the doctor to diagnosis and treatment, a change entailing recognising the undiagnosed patient and recognising no treatment as effective treatment.

ACKNOWLEDGEMENTS

I thank Mr. J. R. Compton, who was responsible for all the statistical work Professor J. A. Forbes, Dr. Ian Skottowe, Dr. Stephen Mackeith, and Dr. D. Mulhall, all of whom gave advice and help.

REFERENCES

1. Fry, J., *British Medical Journal*, 1957, **2**, 1455.
2. Taylor, S., and Clave, S., *Mental Health and Environment*, p. 133. London, Longmans, 1964.
3. Shephard, M., *Psychiatric Illness in General Practice*, p. 95, London, Oxford University Press, 1966.
4. Spencer, D. J., *Journal of the Royal College of General Practitioners*, 1970, **15**, 107.
5. Elliott-Binns, C. P. E., *British Medical Journal*, 1938, **11**, 271.
6. Marinkeer, M., *Journal of the Royal College of General Practitioners*, 1967, **14**, 62.
7. Eimerl, T. S., MD thesis, University of Liverpool, 1962.
8. Crombie, D. L., *Journal of the College of General Practitioners*, 1963, **6**, 582.
9. Research Committee, *Journal of the College of General Practitioners*, 1958, **1**, 107.
10. Thomas, K. B., *British Medical Journal*, 1974, **1**, 626.
11. Balint, M., *The Doctor, His Patient, and the Illness*. Tunbridge Wells, Pitman Medical, 1975.
12. Cartwright, A., *Patients and their Doctors*. London, Routledge and Kegan Paul, 1967.
13. National opinion poll, Association of British Pharmaceutical Industries paper 1972.

14. Stimson, G. V., *Journal of the Royal College of General Practitioners*, 1970, **26**, suppl 1.
15. Stimson, G. V., *Medical Sociology Research Centre*, Occasional papers Swansea, 1975.

6 Measuring quality of care: do good GPs give their patients more time?

Martin Roland

INTRODUCTION

One of the weaknesses in the discipline of general practice is the difficulty of defining and measuring what constitutes good care. As with any discipline, there are a number of components to quality. These include patients having adequate access to care, providing care which identifies and appropriately addresses the needs of patients, and providing care which is acceptable to patients. Within that framework, measuring the outcome of care is an important component. However, outcome measurement, with the exception of the intermediate outcome measure of patient satisfaction, poses particular problems for general practice, as many of the problems we deal with relate to care for patients with multiple long-term, or incurable problems. The focus of most studies which have tried to define the parameters of good quality care has therefore been on issues relating to the process of good general practitioner care. There are a number of papers which relate to one aspect of the process of care, namely the time which a general practitioner spends with his patient. These papers have tried to define the extent to which shortage of time constrains the general practitioner's ability to provide high quality care. The paper discussed in this chapter (Howie *et al.* 1991) is one the more important of this group of papers, reporting the results of a very large study of Scottish general practitioners.

Among the studies which have looked at the relationship between time and quality of care, including Hughes (1983), Morrell *et al.* (1986), Roland *et al.* (1986), Butler and Calnan (1987), and Ridsdale *et al.* (1989), there has been a general conclusion that doctors perform better when they have more time. However, the size of observed differences when doctors have more time has not been very large, and Butler and Calnan (1987) are somewhat sceptical that more time is what general practitioners need to provide better care. Unlike two experimental studies which looked at the effect of giving more time to general practitioners (Morrell *et al.* 1986; Ridsdale *et al.* 1989), the study discussed here is an observational one based on a very large set of data collected by Lothian general practitioners. In a cross-sectional survey data may be collected prospectively or retrospectively, and the analysis will usually involve the exploration of associations between different variables on which data are available. In this study, the belief of the authors, outlined in the introduction, is that 'both patients and doctors feel that the quality of consultation is generally constrained by the shortage of available time', and they have looked for associations between the time spent with patients and a number of measures of process and outcome to provide support for this view. It is important that although an association (for example between time spent and positive outcome of care) may be found in a cross-sectional study, this does not prove that the relationship is causal. The steps which need to be taken in deciding whether a relationship is causal are set out in Section 3 of this chapter.

The paper was chosen because the issue of how much time a general practitioner should spend with his patients seems to me to be a crucial one, especially in a health service where doctors are rewarded financially for having large lists of patients, and therefore inevitably spending less time with them. If this is to be a trend for the future, then we need to know what the cost will be. Studies looking at the link between quality of care and time available for patients are needed to inform debate with government about what an appropriate list size for a general practitioner should be. The paper selected makes an important contribution to this debate. It is a very carefully executed study from one of the outstanding Departments of General Practice in the country. I wish to discuss the range of problems raised in the design of such a study to illustrate how difficult it can be to answer some of the basic questions about our discipline.

The discussion of the paper has been divided into a number of sections:

1. How sound is the study design and analysis?
2. Are there convincing associations between length of time spent with patients and quality of care?
3. Are the associations sufficiently large to be clinically important, and are they likely to be causal?
4. Is the new measure of 'long to short consultation ratio' an improvement on previously used measures, which relate to the average length of time spent with patients?
5. What are the implications of the study for future practice?

When reading a paper on a large and complex study, it is important to realize that journals such as the *British Medical Journal* and the *British Journal of General Practice* have a strict limit on the length of articles that they will publish. It is therefore often not possible to include all details of a study in one paper. It is common practice, as in the case of Howie's paper, for some details, especially finer details of the methods used to be published elsewhere.

1. HOW SOUND IS THE STUDY DESIGN
AND ANALYSIS?

In this study, data were collected from doctors who recorded information on consultations on one day in fifteen over a year. Eighty-six general practitioners were originally recruited into the study, representing 17 per cent of Lothian general practitioners. All the region's doctors had originally been asked to be involved, so this is not a random sample, and represents doctors who were prepared to devote a fair amount of time to collecting research data. It is difficult to know whether this introduces a serious bias into the study. The results of the study can only be regarded as generally applicable if the doctors involved in this study were a representative sample of all doctors — which they were almost certainly not. However, it would be almost impossible to recruit a representative sample of doctors to collect this type of data, so the question is whether the sample selection invalidates the conclusions. The authors have limited data on the representativeness of their sample — they included young and old doctors, male and female, and doctors from small and large practices but they were not able to look for the representativeness of their doctors in terms of the most important characteristic, that is their use of time. It would be a source of concern if doctors agreeing to participate in this type of study were ones who spent longer than average with their patients or those who had relatively small lists of patients, since this would limit the generalizability of the results. We are not given any information in this paper about the list sizes of participating practices compared to other Lothian doctors, but the results show that most frequent time spent with patients was four minutes for the faster consulting group, so rapid consulters appear well represented in the doctors who were studied.

Relatively few checks to the accuracy of the data are described in this paper. The doctors were responsible for timing each consultation. The authors state that the method of timing is described in another paper (Porter *et al.* 1985), but while this earlier paper describes problems of recording interruptions and distractions, it does not say how they were overcome. We do not therefore know whether a 3 minute consultation interrupted by a 10 minute phone call was recorded as 3 minutes or 13 minutes. The doctors may have varied considerably in the accuracy of their time recording. Likewise, there is no description of the accuracy of recording of other information by the doctors, or the extent to which they agreed about definitions when recording 'other health problem', or 'health promotion', the latter having been recorded on a five point scale. Reliable data are very difficult to collect in the course of routine surgery consultations — even by the most committed — and additional validation of the data, for example by the use of audiotapes or independent timing of a sample of consultations, would have strengthened the study, though this would have greatly increased the costs of the study.

Patient satisfaction questionnaires were given to patients by the doctors at the end of each consultation. This method is likely to bias the response given towards satisfaction, particularly if the patients believe that the doctor may see their responses. The trade-off here is against response rate, which might well have been reduced if another method of handing out the questionnaires had been used. Only 43 out of the original 85 doctors agreed to administer the patient satisfaction questionnaire, introducing another potential source of bias, although the direction in this case is not possible to assess.

Analysis

Following collection of the data, doctors were classed as 'fast', 'slow', or 'intermediate' on the basis of their mean recorded consultation times. All doctors had some consultations which were very short, and some which were much longer. The distribution of consultation times (Fig. 1) was very skewed and it would have been more conventional to use the median as the measure of average consultation time rather than the mean for this type of distribution. When distributions are very skewed, data are sometimes transformed (for example by using the logarithm of the values) in order to approximate the values to a normal distribution: this allows a wider range of statistical tests to be used.

2. ARE THERE CONVINCING ASSOCIATIONS BETWEEN LENGTH OF TIME SPENT WITH PATIENTS AND QUALITY OF CARE?

The main associations demonstrated in the paper are:

(a) between consultation length and

(i)   recognition of concurrent problem;

(ii) management of concurrent problem;

(iii) management of relevant psychosocial problems;

(iv) quantity of health promotion;

(v)   patient satisfaction; and

(b) between category of doctor (fast/slow/intermediate) and

(i)   prescribing rate;

(ii) doctor satisfaction.

The question as to whether these are true associations depends principally on whether any of the potential sources of bias described in the previous section could have produced spurious associations.

The first four associations depend on judgements made by the doctors at the time they recorded the length of a consultation. There could have been an effect of the doctors justifying the time that they had spent with a patient, or (worse) relying on memory and over-estimating the time of complicated consultations if they had not always remembered to record the length of a consultation each time a patient left. On the question of patient satisfaction, patients might have been more reluctant to criticize a doctor who had just spent a long time with them, resulting in artificially high satisfaction scores for longer consultations. The sixth association depends critically on the case mix of patients seen by the three groups of doctors — there is too little information on the diagnostic information given by the doctors or its analysis to form a judgement about the importance of this potential source of bias.

There is no answer to the question of how important these potential sources of bias were; this to some extent remains a matter of judgement for the reader. There is no question that in a very large study of this type, it would have been extremely expensive, to allow for these potential sources of bias, and the number of doctors willing to co-operate in the study might have been drastically reduced if a series of checks on validity had been built in. The most serious of the problems outlined above is probably the decision to allow the doctors themselves to give the satisfaction questionnaires to patients. This could have resulted in serious distortion of their views. An alternative would have been for a researcher or receptionist to have given the patient the satisfaction questionnaire as they left the consulting room (with appropriate assurances of anonymity) or for the research team to have mailed on the questionnaire (though this would seriously reduce the response rate).

3. ARE THE ASSOCIATIONS SUFFICIENTLY LARGE TO BE CLINICALLY IMPORTANT, AND ARE THEY LIKELY TO BE CAUSAL?

As the authors point out in the discussion, statistical significance and clinical importance are not the same thing. Particularly in a very large study, quite small differences may be statistically significant even though they are not of great clinical importance. One of the problems for me in reading this paper is the lack of numeric data, much of which can only be gleaned by eye from the figures. For example in Figure 3 there appear to be very large differences in the proportion of psychosocial problems dealt with on long compared to short consultations (approximately 50 per cent compared with 20 per cent of consultations). The differences in reported health promotion behaviour appear to be much smaller, but it is hard to gauge the figures accurately from Figure 4. The authors listed chi-squared values for some of their significance tests in Appendix 1: to have reported mean values and confidence intervals of the variables which they measured would have helped the reader to understand the sizes, of effects which they were able to demonstrate.

For some of the patient satisfaction items, statistically significant differences comparing long with short consultations are clearly large and important (for example 83 per cent vs. 68 per cent of patients reporting that the GP gave them a chance to say what was really on their mind), whereas for others the differences were small (for example 3.2 per cent vs. 5.8 per cent who reported insufficient time to talk). For the doctors' satisfaction with the consultation, no figures are given, nor clinical significance indicated to substantiate the statement that faster doctors reported themselves to have been less satisfied than slower ones.

The differences in prescribing rates appear to be substantial, with slow doctors prescribing 10 per cent less frequently than fast doctors. However, this difference is difficult to interpret without more information on prescriptions issued, more detailed information on case mix, and information on recall rates in the two groups of doctors.

The question of whether the associations demonstrated are actually causal, that is are the changes due to differences in time management or some other interacting variable, can be considered under the headings introduced in Chapters 7 and 8 on cohort and case-control studies. It is important to realize that no study, particularly a cross-sectional one, can prove a causal relationship between two observed events. However, in trying to decide whether the results of a study suggest a causal relationship, a number of questions can be considered.

1. Was there important bias in the study? Possibly, considered in the section above.

2. Are the results plausible? The associations are certainly plausible. However, the direction of causation is not so clear. For example it may be that psychosocial problems are more likely to be unearthed if doctors consult for longer. Equally, it could be that some doctors are good at detecting psychosocial problems, and that consultations in which these are revealed simply take longer.

3. Are the results consistent with other studies? Yes, they certainly appear to be consistent with previous work carried out in this area.

4. How strong are the associations? A causal relationship is somewhat more likely (though by no means guaranteed) when there is a very strong relationship between two variables. For some of the variables, for example the percentage of patients who felt they understood their problem (41 per cent of patients after short consultations, 57 per cent after long

ones), the differences found in this study were quite large, which would be consistent, though certainly not conclusive of a causal relationship between them. In many studies (though not this one), the strength of an association is reported as a correlation coefficient. If the correlation coefficient is squared this gives the proportion of the variation that is explained by the independent variable. For example if there were a correlation coefficient of 0.4 between the stress reported by GPs and the number of night visits carried out in the previous month, this would mean that 0.16 (i.e. 16 per cent) of the variation in stress levels could be explained in terms of night visits carried out.

5. Are there consistent trends in the effects seen with consultation length? Since three lengths of consultation and three types of doctor are being considered, one ought to be able to see trends across the groups and not just differences between the two extreme groups. This is clearly the case for the proportion of psychosocial problems dealt with, where intermediate length consultations appear to identify more problems than short ones and long consultations identify still more. However, this trend is much less clear for health promotion, which seems to be increased in medium length compared to short consultations, but increased no further in long consultations. This is consistent with Morrell's study where consultations booked at 10 minute intervals contained very little more health promotion activity than those booked at 7.5 minute intervals, but more than those booked at 5 minute intervals.

4. IS THE NEW MEASURE OF 'LONG TO SHORT CONSULTATION RATIO' AN IMPROVEMENT ON PREVIOUSLY USED MEASURES WHICH RELATED TO THE AVERAGE LENGTH OF TIME SPENT WITH PATIENTS?

Previous work in this area has used 'length of consultation' as the variable to 'explain' differences in doctors' behaviour and the outcome of their consultations, or has characterized doctors in terms of their average consultation time. Here we are presented with a new concept, the 'long to short consultation ratio', which is the ratio between the number of consultations lasting ten minutes or more to the number lasting five minutes or less. The authors propose this as a measure of doctor style (as opposed to speed) which could be a useful marker for quality of care. The question is whether it improves on previous measures of time spent in consultation.

The authors have produced a measure which is somewhat less easy to understand than a conventional average (e.g. median) consultation time. So they need to present evidence that their new measure is superior. Very little space in the paper is given over to making a case for the new measure, and the authors' case is weakened by the fact that they report most of their results in terms of consultation length rather than consultation ratio. Indeed consultation length has the main effect on long term problems detected, on psychosocial problems detected, and on health promotion activity with relatively little effect of consultation ratio within any one band of mean consultation length. The patient satisfaction data is only reported in terms of consultation length.

Where the new measure does appear particularly valuable is in the data presented in Table 3 which shows that faster and slower doctors have consistent changes in their consulting patterns under a variety of administrative conditions. Given the clear message of the title of the paper, it is perhaps surprising that the authors did not take more space in arguing that it is the doctors' overall style rather than the time spent with an individual patient which affects the outcome of the consultation.

CONCLUSION

I chose this paper as an important paper in the literature investigating general practitioners' use of time, and the relationship between consultation time and quality of care. It is a significant addition to the literature that demonstrates that patients benefit from longer consultations with their doctors.

The paper is a very good illustration of how hard it is to address issues of this type. Part of the difficulty is that the authors rely to a very great extent on the doctor's perception of his own activity during the consultation. There are more objective measures which can be used to collect data of this type. For example, Morrell *et al.* (1986) tape recorded 623 consultations, and used these to count statements relevant to health promotion. In another example, Raynes and Cairns (1980) used an observer in the consultation to record problems detected by the doctor. These methods can clearly only be used on a very small sample of doctors, and they are then open to the accusation of serious selection bias.

Howie and colleagues have taken a different approach, namely to collect a very large volume of data on a large number of general practitioners who are probably a more representative sample of GPs than those included in the studies carried out by Morrell *et al.* (1986) or Rayne and Cairns (1980). However, an inevitable consequence of this is that the authors have less control over the quality of the data collected. Both methods are valid approaches to the problem, and both methods have their associated problems. This is why, in judging the results of any study, it is important to consider how the authors' conclusions fit in with those of other studies, ideally ones using different research methods.

In this case, despite the difficulties associated with the method chosen, the results of the study fit well with previous work done under different conditions, and therefore lend substantial support to the argument that shortage of time constrains the quality of care which general practitioners give to their patients. There remains the question of how the results and conclusions drawn in the study can be used to inform public policy. Here the magnitude of the observed effects become important. Certainly the consultations of 'slow' GPs appear to be associated with some benefits compared with those of 'fast' GPs, including possible reductions in prescribing costs. However, the cost in time to these GPs is considerable. It appears from Figure 1 that the median consultation time is about 5 minutes for 'fast' GPs, and 8-9 minutes for 'slow' GPs. This is a substantial difference in consultation time which has major implications for the average list size which can be sustained. The results of this study provide valuable data on the potential benefits and costs of longer consultations in general practice.

REFERENCES

Butler, J., and Calnan, M. (1987). *Too many patients? A study of the economy of time and standards of care in general practice*. Avebury Press, Aldershot.

Howie, J. G. R., Porter, A. M. D., Heaney, D. J., and Hopton, J. L. (1991). Long to short consultation ratio: a proxy measure of quality of care for general practice. *British Journal of General Practice*, **41**, 48-54.

Hughes, D. (1983). Consultation length and outcome in two group general practices. *Journal of the Royal college of General Practitioners*, **33**, 143-7.

Morrell, D. C., Evans, M. E., Morris, R. W., and Roland, M. O. (1986). The 'five minute' consultation: effect of time constraint on clinical content and patient satisfaction. *British Medical Journal*, **292**, 870-3.

Porter, A. M. D., Howie, J. G. R., and Levinson, A. (1985). Measurement of stress as it affects the work of the general practitioner. *Family Practice*, **2**, 136-46.

Raynes, N. V., and Cairns, V. (1980). Factors contributing to the length of general practice consultations. *Journal of the Royal college of General Practitioners*, **30**, 496-8.

Ridsdale, L., Carruthers, M., Morris, R., and Ridsdale, J. (1989). Study of the effect of time availability on the consultation. *Journal of the Royal college of General Practitioners*, **39**, 488-91.

Roland, M. O., Bartholomew, J., Courtenay, M. J. F., Morris, R. W., and Morrell, D. C. (1986). The 'five minute' consultation: effect of time constraint on verbal communication. *British Medical Journal*, **292**, 874-6.

Long to short consultation ratio: a proxy measure of quality of care for general practice

J. G. R. Howie, A. M. D. Porter, D. J. Heaney and J. L. Hopton
The Department of General Practice, University of Edinburgh
British Journal of General Practice, **41**, 48–54 (1991)

Abstract

Eighty five general practitioners in the Lothian region recorded information on all surgery consultations on one day in every 15 for a year. On the basis of their mean consultation times with patients the working styles of the general practitioners were described as 'faster' (n = 24), 'in-termediate' (n = 40) or 'slower' (n = 21). The 21 707 consultations which they carried out over this period were defined as 'short' (five minutes or less), 'medium' (six to nine minutes) or 'long' (10 minutes or more). Independent of doctor style, 'long' consultations as against 'short' consultations were associated with the doctor: (1) dealing with more of the psychosocial problems which had been recognized and were relevant to the patient's care; (2) dealing with more of the long term health problems which had been recognized as relevant, and (3) carrying out more health promotion in the consultation. Patients also reported greater satisfaction with longer consultations. The ratio of long:short consultations was found to be 0.28:1 for 'faster' doctors as against 2.3:1 for 'slower' doctors. When doctors in either group had more heavily booked surgeries or were running late, the long:short consultation ratio fell, in some cases by over 50%.

This paper suggests that the ratio of long to short consultation length for a general practitioner might become the basis of a simply proxy measure of quality of care; and that its use might help monitor the effect of recent and proposed changes in the way in which general practice care is delivered.

INTRODUCTION

Much of the recent work on the quality of care in general practice in the UK has centred on the availability and use of time at consultations. In 1987, Butler and Calnan[1] discussed recent literature relating list size to quality of care and described a postal study in which 1419 doctors gave information on their use of time both generally and in relation to consultations within their surgeries. Wilkin and Metcalfe[2] in a large study of urban general practitioners found that doctors with smaller lists had longer mean consulting times and Fleming and colleagues[3] have shown benefits in the processes of care in relation to list size and consulting time. Howie and colleagues[4] in a study of the work of 85 doctors in the Lothian region, have reported that doctors with longer mean consultation times explore the psychosocial element of respiratory illness consultations more and prescribe fewer antibiotics than do doctors with shorter consultation times, but the same group[5] have described higher stress levels, associated with running behind schedule, in these 'slower' doctors. Morrell and colleagues[6] in a study in a single practice, found that when doctors offered longer consultation times, more problems were identified and patients were more satisfied with the con-sultation. Ridsdale[7] has repeated this work in two further practices and reported a range of process and outcome benefits with longer booking intervals.

Variation between and within general practitioners in their working styles is a reality, yet some of the causes and effects of the variation are easier to describe and evaluate than others. Work which relates aggregated consulting times (whole sessions divided by the number of patients seen) to clinical processes and outcomes lacks the power of studies where individual consulting times within complete surgery sessions can be studied. The aim of the work presented in this paper was the further exploration of the relationship between quantity and quality of care; the authors' belief was that there is only a weak direct link between the number of patients which a general practitioner sees and the goodness of service they receive, but that there is a stronger link between the length of consultations they receive and the quality of care that follows. Working on the assumption that both doctors and patients feel that the quality of consultations is generally constrained by shortage of available time it was hypothesized that preferred patterns of care may be put at risk by working conditions in which the number of longer consultations falls and the number of shorter consultations rises.

METHOD

Data collection

All 496 general practitioners in Lothian were invited to one of three meetings at which the aims and methods of the study were explained. All interested doctors were subsequently visited at their practices. The 86 general practitioners (17% of all Lothian general practitioners) who were recruited into the study were not randomly selected but represented a cross-section of single-handed doctors and doctors in large group practices, male and female doctors and younger and older doctors. During the first three months of the study, one male and one female doctor withdrew from the study, and one single handed male doctor joined the study leaving 85 general practitioners who completed the project.

The doctors recorded information on all surgery consultations on one day in every 15 for a year from November 1987. A Monday recording day was followed progressively by a Tuesday, a Wednesday, and so on, including Saturdays and Sundays. The arrival time of patients was noted by the reception staff using a digital clock synchronized with a clock in the doctors' consulting rooms. When relevant, the appointment time of patients was also recorded. When each patient entered the doctor's consulting room, the doctor noted the time that the patient came in, and at the end of the consultation noted the time that the patient left. This procedure for timing patients through surgery sessions using synchronized clocks in the reception area and in the doctors' consulting rooms was piloted in a previous study by Porter, Howie and Levinson.[8]

At the end of the consultation, the doctors also recorded details of patients' presenting and other health problems, their clinical management, and doctor satisfaction which was measured on a five point Lickert scale from very dissatisfied (1) to very satisfied (5). This method for collecting consultation data had been piloted in one Edinburgh practice of five doctors over a three week period, and found to be acceptable to the doctors, though it did take 30–45 seconds to complete.

During the latter five months of the study, 43 of the 85 doctors agreed to issue a health status measure (the Nottingham health profile)[9, 10] and a 33-item questionnaire to patients attending for consultations. When patients arrived at the doctor's surgery they were asked to complete the first questionnaire, which included the Nottingham health profile, while they were waiting to see the doctor. As they were leaving the doctor's consulting room, patients were asked by the doctor to complete a satisfaction questionnaire before they left the premises. All questions, other than the Nottingham health profile, were derived from pilot studies in one Edinburgh practice, but had not been further validated.

Definition of variables

The consulting style of the doctors was defined after calculating the mean face-to-face length for all surgery consultations (excluding special clinics, for example, antenatal or child care clinics). The 24 general practitioners with the fastest times (6.99 minutes or less per patient) were described as 'faster' doctors; the 21 doctors in the quartile with the longest times (9.00 minutes or more per patient) were described as 'slower' doctors; and the remaining 40 doctors (7.00 to 8.99 minutes per patient) were described as 'intermediate' doctors. The proportion of faster, intermediate and slower doctors was similar for the doctors who agreed to participate in the patient health and satisfaction study.

Within each of the three doctor styles, the percentage distribution of consultation lengths was displayed and the ratio of long:short consultations calculated. The case mix of patients seen by faster, intermediate and slower doctors was assessed in several ways: using the Nottingham health profile, doctors' diagnostic statements, the age and sex of patients and the proportions of new to return consultations.

Quality of care was defined using three process variables and one outcome measure. For the first process variable the general practitioner was asked to note whether a long term health problem had been recognized at the consultation and, if so, whether it had been dealt with. For the second process variable the general practitioner was asked to note whether the patient had a psychosocial problem which was relevant to his or her care and, if so, whether an attempt had been made to deal with it. The third process variable was the amount of health promotion in the consultation; the doctor was asked to score this on a scale from 1 ('none') to 5 ('a lot'). The outcome measure was the 33-item patient satisfaction questionnaire described above.

Finally, the influence of administrative circumstances on the ratio of long:short consultation lengths was examined by recalculating the ratios of consultation lengths, after controlling for the general practitioner's consulting style, for sessions which ran more than half an hour late as against on time, for sessions which included 15 or more as against nine or fewer patients and sessions with and without booked appointments.

RESULTS

Information was recorded on 21 707 consultations.

Consultation length

For the 24 faster doctors, 15.2% of consultations lasted 10 minutes or more, and 54.1% lasted five minutes or less, giving a ratio of long:short consultation lengths of 0.28:1 (Figure 1a). For intermediate doctors the figures were 35.7%

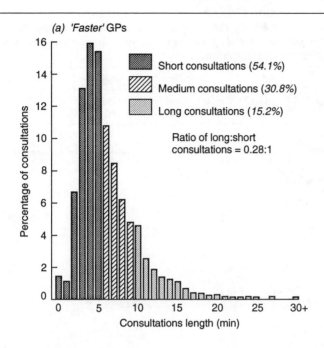

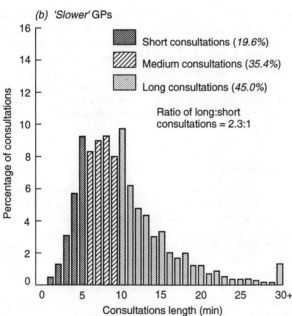

Figure 1. *Distribution of consultation lengths for (a) general practitioners with 'faster' consulting style (*n = 6858 *consultations) and (b) general practitioners with 'slower' consulting style (*n = 4460 *consultations).*

Table 1. *Mean scores on the six dimensions of the Nottingham health profile (NHP) for patients aged 16 years and over prior to consultation according to general practitioner's consulting style.*

Faster practitioner's consulting style	Total no. of patients seen	Patients' mean scores on NHP dimensions					
		Energy	Emotional reaction	Social isolation	Sleep	Pain	Physical mobility
Faster	601	20.39	13.44	5.99	17.92	10.32	5.93
Intermediate	879	21.46	12.89	6.94	20.57	12.13	7.06
Slower	499	20.27	12.65	6.93	16.59	8.51	5.60
Kruskal-Wallis one-way analysis of variance		NS	NS	NS	NS	$P < 0.05$	NS

NS = not significant.

and 26.6%, giving a ratio of 0.75:1. For the slower doctors the figures were 45.0% and 19.6%, giving a ratio of 2.3:1 (Figure 1b).

Case mix

Table 1 shows the Nottingham health profile scores prior to consultation for patients aged 16 years and over. This shows broad similarity in the quantity of physical and emotional morbidity presented to general practitioners in each consulting style group. The diagnostic statements recorded by the three groups of general practitioners were also broadly similar, as were the ages of the patients seen, the proportion of new and return consultations and the number of extra patients fitted into surgery sessions. Slower doctors saw a small excess of female patients compared with faster doctors but not sufficient to explain the observed variations in consultation lengths.

Content of consultations

Figure 2 shows the extent to which concurrent health problems were recognized and explored in short (five minutes or less), medium (six to nine minutes) and long (10 minutes or more) consultations for faster, intermediate and slower doctors (see Appendix 1 for chi-squared values). The chances that long term health problems which had been recognized would be dealt with increased progressively with consultation length, irrespective of doctor's consulting style. Figure 3 demonstrates that the chances that a psychosocial problem which had been recognized would be dealt with also increased progressively with consultation length and irrespective of doctor style. Figure 4 demonstrates the same trends for the amount of health promotion activities.

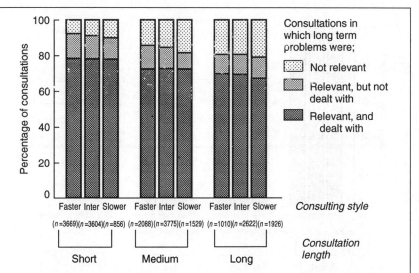

Figure 2. *Percentage of consultations in which long term health problems were dealt with, according to consultation length and general practitioner's consulting style.*

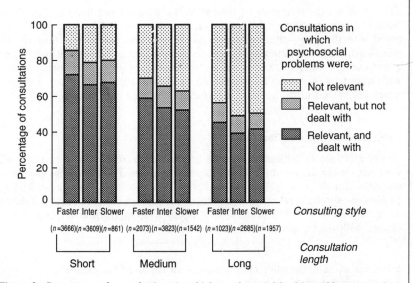

Figure 3. *Percentage of consultations in which psychosocial health problems were dealt with, according to consultation length and general practitioner's consulting style.*

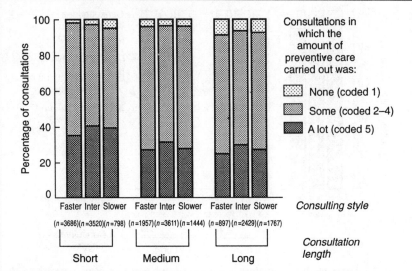

Figure 4. *Percentage of consultations in which preventive care was carried out, according to consultation length and general practitioner's consulting style.*

Patient satisfaction

Table 2 shows the responses to the patient satisfaction questionnaire. A significantly higher proportion of favourable responses were shown for long compared with short consultations for 17 of 33 questions asked ($P < 0.01$ for 11 questions). When the data were reanalysed for the four questions which included a 'not relevant' category, the question 'Did the doctor give you a chance to say what was really on your mind?' remained significant, but the other two questions which are shown in the table as significant, lost their significance.

Other process/outcome measures

The proportions of patients who were brought back, referred and investigated were similar for doctors of different consulting styles. Faster doctors gave prescriptions for drugs at 60% of consultations and slower doctors prescribed at 54% of consultations ($P < 0.01$), and this is in line with our previously published work on antibiotic prescribing.[4] Faster doctors reported lower levels of satisfaction with consultations than slower doctors.

Influence of administrative circumstances on ratios of long:short consultation length

Table 3 shows that the ratio of long:short consultations changed with all the changes in administrative circumstances irrespective of the general practitioner's consulting style.

Table 2. *Results of the satisfaction questionnaire.*

		Percentage of respondents		
		Short consultations [a] (n = 585–721)	Long consultations [b] (n = 579–722)	Significance (χ^2 test)
Consulted this doctor before?	Yes	89.9	87.6	NS
Waiting time was too long?	Yes	14.4	13.4	NS
GP annoyed you?	Yes	1.7	1.1	NS
GP was rather inconsiderate?	Yes	3.1	1.9	NS
GP upset you because it was bad news?	Yes	1.0	1.4	NS
GP upset you because of how he spoke?	Yes	0.4	0.3	NS
GP was difficult to understand?	Yes	0.9	1.3	NS
GP was friendly?	Yes	99.0	99.7	NS
GP was in hurry?	Yes	12.3	6.2	$P < 0.01$
GP was tense?	Yes	3.2	0.9	$P < 0.05$
GP was relaxed?	Yes	92.7	95.5	$P < 0.05$
GP was caring?	Yes	95.1	98.0	$P < 0.05$
GP had other things on his mind?	Yes	3.8	3.2	NS
GP was interested in listening to you?	Yes	96.7	98.5	NS
GP gave you the feeling your opinions were important?	Yes	88.9	94.7	$P < 0.01$
Felt you had plenty of time to talk?	Yes	94.2	96.8	$P < 0.05$
GP took you seriously?	Yes	95.4	96.7	NS
Anything else you would have talked to GP about?	Yes	10.7	13.1	NS
Anything about consultation which disappointed you?	Yes	6.4	5.6	NS
Visit was worthwhile?	Yes	96.9	96.8	NS
Would recommend this doctor?	Yes	97.4	97.9	NS
GP gave you a chance to say what was on your mind?	Yes	67.1	83.4	$P < 0.01$
	No	3.1	1.3	
	Not relevant	29.8	15.3	

Table 2. (*Cont.*).

		Percentage of respondents		
		Short consultations[a] ($n = 585–721$)	Long consultations[b] ($n = 579–722$)	Significance (χ^2 test)
GP helped to reduce your worries?	Yes	54.3	68.2	$P < 0.01$
	No	10.9	9.6	
	Not relevant	34.8	22.3	
GP explained things fully?	Yes	83.5	91.4	$P < 0.01$
	No	2.7	1.9	
	Not relevant	2.7	1.9	
You will follow the treatment?	Yes	91.2	92.1	NS
	No	1.1	0.8	
	Not relevant	7.6	7.0	
Ability to cope with life?	More able	22.9	32.2	$P < 0.001$
	Less able	0.0	0.4	
	Same	77.1	67.3	
Ability to understand illness/problems?	More able	41.2	57.1	$P < 0.01$
	Less able	0.3	0.7	
	Same	58.5	42.2	
Ability to cope with illness/problems?	More able	37.8	53.3	$P < 0.01$
	Less able	0.3	0.4	
	Same	62.3	46.3	
Ability to help self?	More able	35.2	45.7	$P < 0.01$
	Less able	0.2	0.6	
	Same	64.7	53.7	
Confidence about health?	More conf	39.0	46.5	$P < 0.05$
	Less conf	0.7	1.6	
	Same	60.3	51.9	

Ability to get/keep healthy?	More able	33.9	45.5	$P < 0.01$
	Less able	0.2	0.6	
	Same	65.9	53.9	
Feeling after visited the doctor?	Better	50.7	61.1	$P < 0.01$
	Worse	0.4	0.7	
	Same	48.7	37.9	
Satisfaction with consultation?	Very satisfied	59.7	67.2	$P < 0.05$
	Satisfied	33.2	28.6	
	50/50	5.1	2.5	
	Dissatisfied	0.7	0.7	
	Very dissatisfied	1.2	1.0	

n = range of total number of respondents. NS = not significant. [a] Consultations lasting five minutes or less. [b] Consultations lasting 10 minutes or more.

Table 3. *Ratio of long to short consultations according to general practitioner's consulting style for different administrative circumstances.*

	Ratio of long to short consultations (total no. of consultations)		
	Faster GPs	Slower GPs	All GPs
Overall	0.28 (6858)	2.30 (4460)	0.70 (21707)
Appointment sessions	0.31 (5828)	2.41 (3936)	0.75 (18957)
Non-appointment sessions	0.15 (1020)	1.72 (524)	0.45 (2750)
<10 patients per session	0.36 (1151)	3.13 (1058)	0.96 (4623)
>14 patients per session	0.23 (3349)	1.46 (825)	0.44 (6849)
Sessions running on time	0.33 (3965)	2.69 (1752)	0.74 (11605)
Sessions running 15–30 minutes late	0.26 (1437)	2.39 (1275)	0.68 (5274)
Sessions running >30 minutes late	0.20 (1456)	1.85 (1433)	0.63 (4828)

DISCUSSION

Even in a study of over 20 000 consultations, where attempts were made to control simultaneously for more than a few of the many process or contextual variables about which we had information, the numbers available for comparison fell quickly. In addition, many variables were interdependent to some degree and our analyses and interpretations of our data have to be made against these difficulties. At the same time, where the numbers available are large the risk of meaningless statistical significance being found rises. Readers have to judge the balance between statistical and clinical importance for themselves and we have tried to help this by producing detailed tables. However, there are some clear trends.

As would be expected, the content of longer and shorter consultations was different. The greatest differences were first in the number of psychosocial problems identified and dealt with and second in the number of other health problems which were identified and dealt with but were not the primary reason for the consultation. Longer consultations did include more health promotion, but not a great deal more. What was of interest was that doctors with different styles of consulting (faster or slower) appeared to behave similarly when they worked in the same consulting style: for example, when slower doctors worked fast their consultation behaviour appeared to be similar to the behaviour of faster doctors. Long consultations were associated with greater patient satisfaction in several important areas. However, the relationship between patients' expectations of consultations, their perception of the 'relevance' of the questions asked, and satisfaction was not answerable from this data.

Because 'faster' and 'slower' doctors were defined by their mean consultation lengths, it is not surprising that faster doctors had more short consultations and slower doctors more long consultations. However, the evidence we have available suggests that the case mix they see is similar; the diagnostic labels used, the Nottingham health status scores of patients, the proportion of new and return consultations, the age and sex distribution of patients and the extra patients seen, are unable to account for doctors of different style working with the different mix of shorter and longer consultations we have described.

The shape of the distribution curves shows that slower doctors have a much longer 'tail' of long consultations (17% of consultations lasting 15 minutes or more as against 30%). Thus slower doctors are not just faster doctors who work more slowly. Faster doctors had a larger personal list size (mean 1789; standard deviation 775) than did slower doctors (mean 1567; standard deviation 726), but as in our previous paper[4] this difference disappears when partnership totals are calculated as complete units.

Further work will explore the personal and biographical features that might contribute to how doctors adopt the styles they use. In the meantime we recognize that doctors generally feel constrained by their commitments and, although many faster doctors expressed dissatisfaction with short consultations, they did not see a change in organization as a realistic option. Consequently their working pattern becomes relatively fixed and this was reflected in the fact that mean consultation lengths for faster doctors did not vary much with the number of patients they had to see. Slower doctors behaved relatively consistently with sessions of up to 12 patients (about two hours work) but when they had a larger number of patients to see in a session their mean consultation length fell to accommodate the extra patients.

Given that our knowledge of why some doctors are 'fast' and others 'slow' is incomplete, it seems safer to make comparisons between general practitioners using doctors of the same consulting style. Doing this we found that several common administrative problems significantly changed the patterns of time distribution. This happened most noticeably for the slower doctors who started with greater opportunity for flexibility. For them, having a larger number of patients to see in a session as against a smaller number changed the ratio of long:short consultations from 3.13:1 to 1.46:1. The greatest loss was in the number of consultations lasting 15 minutes or more, which fell from 22% to 12% of all consultations while the number of consultations of five minutes or less arose from 16% to 26%. The same trends were seen when doctors were running late and when they worked with unbooked rather than booked appointments. It is worth commenting, however, that the single feature which most commonly correlates with running late is having booking intervals that are incompatible with consultation times (manuscript in preparation). This was a much more common problem for slower doctors, and their patients indicated dissatisfaction with the times they often had to wait (manuscript in preparation).

Given the comments we have made about the similarity of the case mix, it is difficult not to conclude that differences in the ratio of long:short consultations reflect differences in the quality of care being offered, certainly within doctor style and possibly between doctor styles. The advantages of longer consultations do not simply lie in more services being provided but in a larger proportion of the needs which have been recognized being followed up by the doctor. It seems reasonable to equate shorter consultations with 'presenting symptom' medicine and longer consultations with the wider interpretation of consultations as described by the Stott and Davies model.[11] There are those who argue that the 'presenting symptom' model is the appropriate one and that the patients do not wish for or benefit from wider discussion of perhaps insoluble problems, but the responses to our satisfaction questionnaire do not support this view. We have yet to investigate the economic implications of these different consulting styles.

At present there is considerable concern that changes in the contracts and modes of working of general practitioners will adversely affect the quality of care they offer. If more time has to be spent on administration or on the provision of new services, this may be reflected either in changes in the time available for consultations, or in how that time is used. Longer consultations (especially those over 15 minutes) may have to be sacrificed as part of a change to shorter average lengths. Alternatively average consultation lengths may arise but the number of longer consultations may fall because of an increase in the time spent recording information.

If longer consultations are better than shorter ones, then our method of displaying the distribution of consultation lengths and trying to interpret the shapes of these distributions may offer a practical and sensible way of monitoring the quality of care delivered in general practice in the years ahead. We recommend its further exploration.

ACKNOWLEDGEMENTS

We are greatly indebted to the 85 general practitioners whose commitment to this project has made this paper possible. We would also like to thank our colleagues in our research team who have helped process the data. The project was supported by the Health Service Research Committee of the SHHD and by the Nuffield Provincial Hospitals Trust. Jane Hopton is supported by the Andrew Robertson Medical Fund.

APPENDIX 1

The following tables show the chi-square values and statistical significance of the differences in the processes of care recorded by general practitioners between length of consultation and between general practitioners' style of consultation. In general, significant differences were found between type of consultation within the same style, rather than between style within type of consultation.

	Long term health problems			
	Not relevant, relevant (not dealt with) and relevant (dealt with)		Relevant (not dealt with) and relevant (dealt with)	
GP consulting style (faster/ inter/slower) controlling for consultation length				
Short	11.4	$P = 0.07$	6.8	$P < 0.05$
Medium	14.0	$P < 0.05$	13.2	$P < 0.001$
Long	26.0	$P < 0.001$	17.6	$P < 0.001$

Consultation length (short/medium/long) controlling for GP consulting style				
Faster	164.7	P < 0.001	45.0	P < 0.001
Intermediate	156.4	P < 0.001	48.1	P < 0.001
Slower	109.0	P < 0.001	34.5	P < 0.001

	Psychosocial problems	
	Not relevant, relevant (not dealt with) and relevant (dealt with)	Relevant (not dealt with) and relevant (dealt with)
GP consulting style (faster/ inter/slower) controlling for consultation length		
Short	45.5 P = 0.001	6.5 P < 0.05
Medium	34.3 P < 0.001	10.2 P < 0.01
Long	15.0 P < 0.05	4.1 P = 0.13
Consultation length (short/medium/long) controlling for GP consulting style		
Faster	594.7 P < 0.001	85.9 P < 0.001
Intermediate	887.9 P < 0.001	158.9 P < 0.001
Slower	328.7 P < 0.001	66.9 P < 0.001

	Preventive care
	Codes 1, 2, 3, 4 and 5
GP consulting style (faster/ inter/slower) controlling for consultation length	
Short	100.0 P = 0.001
Medium	55.6 P < 0.001
Long	73.7 P < 0.001
Consultation length (short/medium/long) controlling for GP consulting style	
Faster	224.4 P < 0.001
Intermediate	196.7 P < 0.001
Slower	62.4 P < 0.001

REFERENCES

1. Butler, J. R., Calnan, M. W. List sizes and use of time in general practice. *Br Med J* 1987; **295:** 1383-1386.
2. Wilkin, D., Metcalfe, D. H. H. List size and patient contact in general medical practice. *Br Med J* 1986; **289:** 1501-1505.
3. Fleming, D. M., Lawrence MSTA, Cross, K. W. List size, screening methods and other characteristics in relation to preventive care. *Br Med J* 1985; **291:** 869-872.
4. Howie, J. G. R., Porter, A. M. D., Forbes, J. F. Quality and the use of time in general practice: widening the discussion. *Br Med J* 1989; **298:** 1008-1010.
5. Porter, A. M. D., Howie, J. G. R., Forbes, J. F. Stress in general medical practitioners of the United Kingdom. In: McGuigan FJ, Syme WE, Wallace JM (eds). *Stress and tension control 3.* London: Plenum Press, 1989.
6. Morrell, D. C., Evans, M. E., Morris, R. W., Roland, M. O. The 'five minute' consultation: effect of time constraint on clinical content and patient satisfaction. *Br Med J* 1986; **292:** 870-873.
7. Ridsdale, L., Carruthers, M., Morris, R., Ridsdale, J. Study of the effect of time availability on the consultation. *J R Coll Gen Pract* 1989; **39:** 488-491.
8. Porter, A. M. D., Howie, J. G. R., Levinson A. Measurement of stress as it affects the work of the general practitioner. *Fam Pract* 1985; **2:** 136-146.
9. Hunt, S. M., McEwen, J., McKenna, S. P. Perceived health: age and sex comparisons in the community. *J Epidemiol Community Health* 1984; **38:** 156-160.
10. Hunt, S. M., McEwan, J., McKenna, S. P. *The Nottingham health profile user's manual.* Manchester: Galen Research and Consultancy, 1989.
11. Stott, N. C. H., Davis, R. H. The exceptional potential in each primary care consultation. *J R Coll Gen Pract* 1979; **29:** 201-205.

7 Exploring out-of-hours demands: a case-control study

Penny Owen, Clare Wilkinson, and Paul Kinnersley

INTRODUCTION

In this chapter we review a case-control study of general practitioners' out-of-hours visits. Case-control studies are a classic method of investigating causes and risks in medicine; a detailed review of this research method has been provided by Hayden *et al.* (1982). In an ideal case-control study the cases are people who are identified as having the disease or outcome that is of interest. The controls are people who do not have the disease, but who are in other respects representative of the population from which the cases arise. General practice age–sex registers can provide a particularly useful tool for obtaining suitable controls. The risk factors or exposures of interest must be carefully defined, and the investigator then retrospectively determines the exposure rate for these risk factors in the cases and the controls.

The results can be displayed in a conventional 2 × 2 table:

		Cases	Controls
	Yes	a	b
Exposure to risk factor			
	No	c	d

Calculation of the odds ratio $\dfrac{(a/c)}{b/d}$ provides a measure of how much more (or less) likely it was for the cases to have been exposed to the risk factor than the controls.

Statistical significance can be determined either by the use of the chi-square test, or by calculating the 95 per cent confidence interval (CI) for the odds ratio. If the 95 per cent CI includes the value of 1.0 then the null hypothesis, that there is no difference between the cases and controls with regard to exposure to the risk factor, cannot be rejected with confidence.

The design of a case-control study is such that sampling of cases occurs by disease or outcome as opposed to risk factor or exposure and the study looks backwards from effect to cause. This is in contrast to randomized controlled trials and cohort studies in which the study groups are defined by risk factor not disease, and then followed forward from possible cause to expected effect. The results of case-control studies have to be interpreted with caution due to the limitations of this study design and it is important that undue value is not placed on the results of a single study. However, they provide a relatively quick and inexpensive method of studying the relationship between cases, particularly rare cases, and a range of possible aetiological factors.

A CASE-CONTROL STUDY OF CHILDREN SEEN FREQUENTLY OUT-OF-HOURS IN ONE GENERAL PRACTICE

Morrison and her colleagues (1991) describe how this study arose from their own observations that during their daily work it felt as if 'a small number of children were generating an excessive amount of out-of-hours work'. Recognizing that 'doctors impressions' cannot always be trusted, they wished to investigate whether there was indeed such a group of children and identify reasons for their high use of the service. This study provides a good example of how general practitioners can turn observations on their everyday clinical work into research. Gathering more information about a part of our work which we find difficult or unsatisfactory in some way can help to change our attitudes to it. Out-of-hours work places a considerable burden on general practitioners, so studies enabling us to understand better why some patients are seen more often than others or even leading to changes in the way care is called for or offered would make an important contribution to practice.

By defining and then studying the children who were generating the majority of the out-of-hours work, the authors hoped to be able to suggest ways of modifying this behaviour. The study design chosen is interesting. It consists of a year long survey to define the cases and explore whether indeed a small number of children were relatively high users of out-of-hours services. Into this survey the case-control study is 'nested'. The relatively small size of the study allowed for home interviews of mothers to collect data on possible factors contributing to service use and this included an attempt to understand mothers' decision making by use of a vignette instrument.

The authors do not provide their own prior hypotheses as to why some children are seen more often than others until the discussion, and the reader is left to infer them from the list of data to be collected under 'Methods'. Both cases and exposures should be clearly defined before embarking on a case-control study, along with the hypotheses describing their relationship. For example Doll and Hill (1950) clearly stated that the purpose of their investigation was to determine whether patients with carcinoma of the lung (the disease/outcome) differed materially from other persons in respect of their smoking habits (risk factor/exposure). In contrast, in this study of children seen frequently out-of-hours a wide range of socio-economic and medical information on the cases and controls is collected, and the differences the researchers found are then described. In fact, the implicit hypotheses are sensible and grounded in the literature, relating to mothers' behaviour, assessment of illness severity, and degree of home support.

The authors identified forty children under the age of 10 years who had received two or more visits in one year. These cases were then matched for age and sex with forty other children from the same practice. Unfortunately telephone contacts were excluded. Since access to a telephone is associated both with affluence and easy medical contact, it could affect interpretation of apparent relationships between affluence and face-to-face contact. The study results showed that the mothers of the cases were more likely to be divorced or single, to have a lower educational level, and to be receiving income support. They were less likely to own their own home or a car. The cases were indeed slightly less likely to have a telephone than the controls. The understandable decision not to collect unreliable telephone contact data prevents any analysis of the differences in behaviour of the cases and the controls which relates to telephone use.

The authors found no difference in the number of serious health problems occurring in the cases and controls. This is rather surprising since a greater level of morbidity would be expected amongst children from more socially disadvantaged families, as the cases are shown to be. As Blaxter found 'families with young children, in disadvantaged circumstances, are always likely to need a great deal of the doctor's time' (Blaxter and Paterson 1982). The authors searched the children's records 'for evidence of physical ill health other than minor self limiting illness' and found five such problems among the study group and six among the controls. Thus 14 per cent of the children had suffered some episode of substantial ill health. It would be helpful to others wanting to compare their own practice with that reported, if the definition of physical ill health had been more explicit. However, while a question-mark must hang over the background morbidity in cases and controls, convincing evidence is presented that out-of-hours calls themselves were mainly for minor illness. It would help the argument if the diagnoses recorded for cases and controls and subsequent action could have been presented separately. Numbers are too small for this to be sensible. In addition, severity was not defined by doctors at each visit. Thus on both counts we cannot be confident that there was no real difference in morbidity between cases and controls to explain the differences in consulting behaviour.

This question of the potential differences in the morbidity of the cases and controls is another example of that particular weakness of the case-control methodology in defining aetiology: namely the possibility of a confounding variable. A confounder is a factor that affects both the outcome and, independently, the risk factor under investigation. In this study it is possible that the confounding variable of high morbidity not only predisposes to the outcome, i.e. high use of the out-of-hours service, but also affects independently a proposed risk factor, for example the degree of anxiety in mothers. The presence of a telephone in a household can also be seen as a potential counfounding factor in this study, since having a telephone may be associated with higher social class and, independently, with a lower call-out rate (through easier access to advice).

The principle finding reported was that while 94 per cent of children under 10 years in the practice received less than two visits annually, 6 per cent of the children received 75 per cent of all visits. These were the children who entered the case-control study as cases and the results of this part of the project showed that they came from more socially disadvantaged families; their mothers were more often single parents and they were more dependent on a range of indices than the control children. These findings do not surprise of themselves. They join a growing literature on out-of-hours work in general practice and its relationship to morbidity and social deprivation (Iliffe and Haug 1991). Most out-of-hours calls are from parents about their children (Wilkin *et al.* 1987); this may reflect parental inexperience (Cunningham-Burley 1990); lack of social support driving parents to seek professional help urgently (Virji 1990), or maladaptive behaviour transmitted as a family tradition (Huygen 1991). Elderly people are the second most frequent callers and are taken more seriously by their general practitioner than are anxious parents (Wilkin *et al.* 1987). Social class seems in some way to act as an important determinant of medical calls (Wilkin *et al.* 1987) but some evidence suggests that working class people. value house visits more than do the middle class and request them less often (Carwright and Anderson 1987). Nevertheless, people living in deprived inner city areas do tend to have high rates of out-of-hours visits (Wilkin *et al.* 1987; Livingstone *et al.* 1989). A study of Family Health Service Authorities demonstrated that lower rates of night visiting were associated with lower standardized mortality rates, i.e. with healthier populations (Baker and Klein 1991). The present study does add the intriguing findings of the vignette responses. Mothers of children who had more out-of-hours visits were more likely to suggest contacting a doctor as the right response to a vignette case than control mothers. They were also more likely to call out a doctor for themselves than the control mothers.

The strength of this study can be seen as the capture of a feeling of overwork and its description by a range of methodologies. A survey of consulting behaviour, a case-control study of consulting determinants, and within this a vignette response study of mothers' decision making. The weaknesses include difficulties with defining all out-of-hours contacts, possible confounding effects e.g. of morbidity definition and telephone use, and the small size of the study. Given these difficulties what can we conclude? It might be argued that the variation in out-of-hours work load amongst this patient group is understandable and to be expected. Some doctors may be content to regard this as part of their job, others may be intrigued by the pointers here and ask is it like this in my practice? What are the mechanisms underlying the observation that a minority of children receive the majority of housecalls? Can we define morbidity and severity better in a future study? Does access to a car and/or telephone really make a significant contribution? Why do some mothers feel that a particular case needs to see a doctor while other mothers are confident without advice?

Further qualitative work regarding the perceptions of patients might be considered. This is because the case-control design in medical research is particularly vulnerable to the preconceived constructs of those who design the study. It may be that patients' reasonable requests for calls to vulnerable children would be best managed by reorganizing our methods of coping with out-of-hours demands. Alternatively, future prospective studies might test the effectiveness of interventions designed to provide social support or health care information (or even subsidized telephones) to isolated or single parent families. The outcome measures would include out-of-hours demands, but might also include other measures such as the Liverpool questionnaire used in this study. Use of a validated instrument enables comparison of results from other studies. Prospective studies of interventions such as extra education would need to be large and well controlled (Tudiver *et al.* 1992).

COMMON PITFALLS IN THE USE OF THE CASE-CONTROL METHODOLOGY

The problem of confounding variables has already been raised in the discussion of the study by Morrison *et al.* (1991). When a variable is known to be related independently to both the risk factor and to the disease or outcome its effect can be reduced by matching the controls to the cases so as to account for the presence or absence of the confounding variable. For example for conditions that increase in prevalence with age the cases and controls can be selected to have similar age distributions to reduce confounder bias. Alternatively, as long as the confounder can be measured, it can be allowed for in the analysis rather than by matching and much of the statistical methodology in epidemiology is concerned with this issue. This strategy is often preferable as it avoids the problem of over-matching cases and controls. However, it is the confounder that has not been thought of or that cannot be measured, which most threatens the interpretation of case-control studies.

Other pitfalls commonly found in case-control studies cannot be excluded in the paper by Morrison *et al.* For example bias in the selection of the cases or the controls may invalidate a case-control study. Over-selection of controls, for example by choosing them from a generally healthier group than the population from which the cases came, may occur. In studies investigating a possible association between reserpine and breast cancer (Boston Collaborative Drug Surveillance Program 1974; Heinonen *et al.* 1974) patients with cardiovascular disease or those who had undergone cardiac surgery were excluded only from the control group of patients without breast cancer. Since these excluded patients were particularly likely to be receiving reserpine, their exclusion would decrease the exposure rate among those chosen as controls and hence falsely elevate the apparent carcinogenic risk of reserpine calculated by comparison with the exposure of the cases.

The importance of diagnostic and disease-staging accuracy amongst those selected as cases and controls is illustrated by the aftermath of the study which prematurely devalued the treatment offered by the Bristol Cancer Help Centre (Bagenal *et al.* 1990). In this study, although there was little obvious difference in the formal prognostic features of the cases and the controls, 42 per cent of the cases had metastatic disease at the start of the study compared with only 29 per cent of the controls (Hayes *et al.* 1990). The cases were also younger and therefore likely to be suffering from a more aggressive disease. The authors of the study finally conceded that the differences they identified between the cases and controls were likely to have been due to increased severity of disease amongst the cases rather than any effect due to attending the Bristol Cancer Help Centre (Chilvers *et al.* 1990). Despite their best attempts, we are left unclear in the study by Morrison *et al.* (1991) as to whether the cases were more seriously ill, or more vulnerable to minor illness, than the controls, and this constrains the interpretation of the results found. The difficulties inherent in choosing cases and controls are such that the advice of an epidemiologist or statistician during the design phase of the study is invaluable.

Another problem with case-control studies is the possibility that recall bias may be introduced by the retrospective collection of information on the exposure or risk factor of interest. Differential recall may occur between cases and controls. For example if the cases are patients who already have a disease they may have given considerable thought to the aetiology of their disease. Therefore they may be more likely to remember a particular exposure. In contrast, the healthy controls may forget details of their medical history. Ideally, interviewers and other data gatherers should be blind as to whether a patient is a case or a control and to the risk factor of interest, to prevent the introduction of bias in the process of data collection.

Once a case-control study has been completed care should be taken not to over-interpret the data and to draw only conclusions which can be fully justified by the results. The cautious interpretation of results made by Vessey *et al.* (1979) in their study suggesting that oral contraception protected against the spread of breast cancer was laudable and allowed a further understanding of this relationship without over-confident conclusions. In contrast to this, the Bristol Cancer Help Centre study results were presented to the national media at a press conference. Only those who followed the subsequent correspondence in the *Lancet* would have been aware of the final interpretation of the results.

An example of a model case-control study was conducted by Doll and Hill (1950) in the late 1940s, to determine whether patients with carcinoma of the lung differed materially from other persons in respect of smoking habits or in some other way which might be related to atmospheric pollution. They paid particular attention to the selection of cases and controls, and possible sources of bias in recording. They concluded that there is a real association between carcinoma of the lung and smoking with, for example, an approximate relative risk of 50 for the likelihood of people over the age of 45 years who smoke 25 or more cigarettes a day developing lung cancer when compared with non-smokers. The results of this case-control study continue to influence clinical practice today.

Relative risks of 50 are not given to the researcher every day, but when carefully designed, case-control studies in general practice could yield similarly important results. This particular research methodology should be used more often in primary care, and Morrison and colleagues are to be congratulated for having the imagination to do so.

REFERENCES

Bagenal, F. S., Easton, D. F., Harris, E., Chilvers, C. E. D., and McElwain, T. J. (1990). Survival of patients with breast cancer attending the Bristol Cancer Help Centre. *Lancet*, **336**, 606-10.

Baker, D., and Klein, R. (1991). Explaining outputs of primary health care: populations and practice factors. *British Medical Journal*, **303**, 225-9.

Blaxter, M., and Paterson, E. (1982). Consulting behaviour in a group of young families. *Journal of the Royal College of General Practitioners*, **32**, 657-62.

Boston Collaborative Drug Surveillance Program (1974). Reserpine and breast cancer. *Lancet*, **ii**, 669-71.

Cartwright, A., and Anderson, R. (1987). General practice revisited. Tavistock, London.

Chilvers, C. E. D., Easton, D. F., Bagenal, F. S., Harris, E., and McElwain, T. J. (1990). Bristol Cancer Help Centre. *Lancet*, **336**, 1186-8.

Cunningham-Burley, S. (1990). Mothers' beliefs about and perceptions of their children's illnesses. In *Readings in medical sociology*. S., Cunningham-Burley and N. P., McKaganey. pp. 85-109 Tavistock/Routeledge London.

Doll, R., and Hill, A. B. (1950). Smoking and carcinoma of the lung. *British Medical Journal*, **ii**, 739-48.

Hayden, G. F., Kramer, M. S., and Horowitz, R. I. (1982). The case-control study:

a practical review for the clinician. *Journal of the American Medical Association*, **247**, 326–31.

Hayes, R. J., Smith, P. G., and Carpenter, L. (1990). Bristol Cancer Help Centre. *Lancet*, **336**, 185.

Heinonen, O. P., Shapiro, S., Tuominen, L., and Turunen, M. I. (1974). Reserpine use in relation to breast cancer. *Lancet*, **ii**, 675–7.

Huygen, F. (1991). Family medicine: the medical life history of families. Royal College of General Practitioners, London.

Iliffe, S., and Haug, U. (1991). Out-of-hours work in general practice. *British Medical Journal*, **302**, 1584–6.

Livingstone, A., Jewell, J. A., and Robson, J. (1989). Twenty four hour care in inner cities: two years out-of-hours workload in East London general practice. *British Medical Journal*, **299**, 368–70.

Morrison, J. M., Gilmour, H., and Sullivan, F. (1991). Children seen frequently out-of-hours in one general practice. *British Medical Journal*, **303**, 1111–14:

Tudiver, F., Bass, M. J., Dunn, E. V., Norton, P., and Stewart, M. (1992). *Assessing interventions: traditional and innovative methods*. Research methods for primary care, Volume 4. Sage Publications, London.

Vessey, M. P., Doll, R., Jones, K., McPherson, K., and Yeates, D. (1979). An epidemiological study of oral contraceptives and breast cancer. *British Medical Journal* **i**, 1755–8.

Virji, A. (1990). Study of patients attending urgently without appointments in an urban general practice. *British Medical Journal*, **301**, 22–6.

Wilkin, D., Hallam, L., Leavey, R., Metcalfe, D. (1987). Anatomy of urban general practice. Tavistock, London.

General practice: children seen frequently out of hours in one general practice

Jillian M. Morrison, Harper Gilmour, Frank Sullivan
Dept. of General Practice, University of Glasgow
BMJ, 303, 1111-4 (1991)

Abstract

Objective—To identify reasons why some children receive more out of hours visits than most.

Design—A one year prospective study to identify the study group. This was followed by a case-control study involving a record search and personal interviews.

Setting—One three doctor urban general practice in West Lothian with 4812 patients.

Subjects—40 children aged under 10 years identified as high users of the out of hours service (more than two visits a year) and 40 age and sex matched controls.

Main outcome measures—Numbers of visits; social factors such as lone motherhood, low educational attainment; score for management response to clinical vignette.

Results—147/756 (19%) out of hours visits in the study year were to children aged under 10 years; 109 (74%) to 41 children (6%). Problems seen were mainly minor, and little active management was required. Family and social factors which were significantly more common for the cases than for the controls included a lone mother (15 v 4), low educational attainment by the mother (25 v 14), receipt of income support (22 v 7), and non-ownership of the home (45 v 22) or a car (19 v 9). Mothers of the cases were more likely to choose to contact a doctor when presented with vignettes describing common childhood illnesses (median score for 16 vignettes 16.5 for cases v 14.5 for controls, Wilcoxon signed rank test, p = 0.01).

Conclusions—Children seen more frequently than expected out of hours came from more socially disadvantaged families and their mothers were more likely to seek medical advice about minor childhood illness. Maternal education, to promote confidence in managing minor illness, may reduce their use of the out of hours service.

INTRODUCTION

Perhaps the most stressful aspect of a general practitioners's workload is out of hours work.[1] Research has suggested that among those receiving house visits out of hours children are overrepresented[2] but that few of the problems encountered are serious enough to merit hospital admission.[3]

Parental anxiety and a lower threshold among doctors for responding to requests for visits to children may mean that out of hours consultations for children will always be more frequent than their numbers in the population might warrant, but the general practitioners in the study thought that a small number of children were generating an excessive amount of the out of hours work. Doctors' impressions cannot always be trusted,[4] so this study aimed at finding out whether there was indeed a group of children who were high users of the out of hours service, identifying reasons for their high use, and suggesting ways of modifying it.

METHODS

The study was carried out in an urban practice of three doctors in West Lothian with a list size of 4812 at the midpoint of the study.

Out of hours visits were defined as all face to face contacts between doctor and patient between 6 pm and 8 am, after 1 pm on a Wednesday, after 12 noon on Saturday for the weekend, and on public holidays. Telephone contacts were excluded as we could not be sure that they were adequately recorded by the participating general practitioners. The out of hours work in this practice was shared with two neighbouring practices so that nine general practitioners (one working with a trainee) looked after about 18 000 patients out of hours.

All out of hours visits performed in this general practice were recorded and analysed for one year from 1 November 1988 to 31 October 1989. The doctor on call completed a standard form which was passed to the practice concerned the next morning. The form included details of the date and time of the consultation; the name, address, and age of the patient; the name of the patient's own general practitioner; the nature of the problem, diagnosis, and action taken; and whether any follow up was required.

Children identified in this study as high users of the service out of hours—that is, receiving two or more visits in one year—were matched with other children in the practice of the same sex and age (within one month). The case records of the two groups were then searched for information about their consultation behaviour (excluding attendances for routine paediatric surveillance and immunisations) and physical health. To check that the controls were not presenting elsewhere with acute health problems, we sought evidence of attendance at the accident and emergency departments from the case records.

The mothers of all of the children were visited at home and interviewed by JM. A structured interview with capacity for explanation and discussion was used. Information about the family, including its social characteristics and the mother's opinion of its health, was obtained. Economic determinants such as home ownership, car ownership, employment status, and receipt of income support were chosen over other definitions of social class because we thought that this information was likely to be more discriminating. The vignette instrument developed by Campion and Gabriel to measure mothers' responses to common childhood illnesses was incorporated in the interview.[5] The 16 vignettes represent a wide range of symptoms and problems occurring in children aged 3 months to 9 years described in lay terms. The mothers were asked to make two judgments about the vignettes: how serious the problem was and what action they would ideally take. The problem was scored very serious (2), fairly serious (1), not serious (0). These scores were summed over the 16 vignettes to give each individual mother a total score. The preferred action was scored 2 if it meant consulting a doctor (in the accident and emergency department, in the surgery, or on a house call), 1 if it involved phoning the health visitor, and 0 if it did not include contact with a health professional.

Finally, the mothers' records were searched to find out about their consultation behaviour (excluding antenatal and postnatal appointments) and physical and mental health.

Statistical analysis was by Wilcoxon signed rank test for comparing medians of numerical variables which were not normally distributed—for example, the responses to the vignettes. The paired *t* test was used for normally distributed variables, such as the age of the mother. McNemar's test was used for comparison of proportions. The Minitab statistical package was used to perform the statistical analysis.[6] Confidence intervals for differences in proportions from paired samples were calculated using the Confidence Intervals Analysis statistics package.[7]

RESULTS

In the study year there were 756 out of hours visits in the practice—a rate of 157/1000. There were 694 children in the practice aged under 10 years (14% of the practice), and 147 of.the out of hours visits were to children (19% of visits). Forty one of the children had two or more visits, and these children received 109 of the 147 visits. Therefore 6% of children received 74% of the out of hours visits to children aged under 10 years. Forty (19 girls and 21 boys) of the 41 children were included in the study as one child moved out of the practice during the study year. The average age of the study group was 38.7 months. The peak age for visits was 6–30 months (see figure), but children in this age group who were seen twice or more often out of hours were still in the upper fifth of this age group.

The problems encountered on the 107 visits to the study group are shown in table I and the management is given in table II. The median number of daytime

Table I *Diagnosis in 107 consultations made out of hours by 40 study children*

Condition		No
Respiratory conditions		56
Upper respiratory tract infections	29	
Lower respiratory tract infections (including bronchiolitis)	3	
Asthma	4	
Croup	2	
Otitis media	12	
Tonsillitis	5	
Blocked nose	1	
Gastrointestinal conditions		20
Diarrhoea and vomiting	14	
Colic	4	
Constipation	1	
Non-specific abdominal pain	1	
Accidents		12
Head injury	8	
Fracture of distal phalanx	1	
Sunburn	1	
Sore foot	1	
Foreign body in ear	1	
Other		19
Rash including exanthemas	7	
Maternal anxiety	4	
Raised temperature—no obvious cause	3	
Epistaxis	1	
Teething	1	
Conjunctivitis	1	
Failure to thrive	1	
Don't know	1	

Table II *Management at 107 out of hours consultations*

Management	No
Reassurance or advice	50
Prescription issued	31
On the spot medication dispensed (antibiotic, paracetamol, or electrolyte solution)	17
Hospital admission arranged	4
Referral for radiography	2
Laceration steristripped	2
Information not available	1

consultations was five for the study group and three for the controls (Wilcoxon signed rank test, p < 0.001). According to the records five of the study group and two of the controls had been taken by their parents to the accident and emergency department during the study year. The accuracy of these figures was checked by comparing the practice records with a computerised printout of practice attenders obtained from the local casualty department. Over five months 187 patients in the practice had referred themselves to the accident and emergency department according to the computer printout and 184 letters were held in the patients' records (98% agreement). Therefore the figures for attendance at the accident and emergency department among the children were likely to be accurate. A search of the children's records for evidence of physical ill health other than minor self limiting illness showed five problems among the study group and six among the controls.

The mean age of the mothers of the study children was 27 years (range 18-36) and of the controls 29 years (range 19-45; paired *t* test, p = 0.11). There was no significant difference in median family size (2.0 for study group *v* 2.0 for controls) or in the position of the index child in the family (23 *v* 18 were the first child; McNemar's test, p = 0.30).

The mothers of the study children were more likely to be single or divorced, to have attained a lower educational level, and to be receiving income support (table III). They were less likely to own their own home or a car. There was no

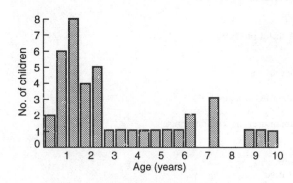

Age range of 40 study children in 6 month bands

Table III *Results of interview with mother about social aspects of family*

	Study group (n = 40) No (%)	Controls (n = 40) No (%)	p Value (McNemar's test)	95% CI for difference between proportions
Single or divorced mother	15 (38)	4 (10)	0.002	9% to 32%
Mother left school with no qualifications	25 (63)	14 (35)	0.02	2% to 46%
Mother employed	21 (53)	18 (45)	0.47	−16% to 28%
Partner employed (based on 26 pairs who both had partners	29 (73)	37 (92)	0.10	−32% to 7%
Receiving income support	22 (55)	7 (18)	0.001	16% to 46%
Own home	5 (13)	18 (45)	0.002	−41% to −12%
Own a car	21 (53)	31 (78)	0.02	−39% to −2%
Have a telephone	25 (63)	33 (83)	0.06	−36% to 3%
Close relatives living nearby*	35 (88)	36 (90)	0.71	−14% to 11%
Goes to nursery, playschool, etc	12 (31)	14 (35)	0.76	−25% to 20%

* Within three miles.

significant difference in whether the mother or her partner was employed, whether there were close relatives living nearby, or whether their preschool children attended a nursery or other preschool group.

There was no difference between the mothers in the perception of whether they thought their children healthy (35 study children and 38 controls were thought to be healthy). There were also no differences in whether they perceived no problems in other family members (32 study group mothers and 36 controls), whether they perceived themselves to be healthy (32 and 31), and whether they perceived no problems in extended family members (31 and 35).

The mothers of the study children were not more likely to consider the vignette situations more serious than the mothers of the controls (median score 12 for cases and 11 for controls, Wilcoxon signed rank test, p = 0.14), but the mothers of the cases were significantly more likely to contact a doctor (Wilcoxon signed rank test, p = 0.01). The median score for the preferred action was 16.5 for the cases and 14.5 for the controls. When asked under what circumstances they would call a doctor out of hours all the mothers said that it would be for an emergency, something they considered very serious, or something they could not cope with, or they described a situation such as a child screaming in pain.

There was no significant difference in the annual daytime consultation rates between the mothers of the study children (6.1) and the mothers of the controls (5.0). There were 17 out of hours visits to 13 of the mothers of the study children in the study year compared with three to the mothers of the controls (McNemar's test, p < 0.01). Only five and three of the mothers respectively attended the accident and emergency department in the year. There was no evidence of a difference in past or present physical or psychological ill health between the two groups of mothers and no difference in their use of psychotropic drugs.

DISCUSSION

The relatively small size of the study group may have resulted in a failure to detect some differences between groups. A sample size of 40 children and 40 matched controls has a roughly 80% power for detecting a difference of 25% between the percentages of cases and controls with some attribute at the 5% level of significance.[8] Using a cut off point of two or more out of hours visits to denote frequent use is contentious, as one child with a serious illness might have received two visits in the course of that illness. However, as 94% of children received fewer than two visits (and 89% were not visited at all out of hours) we thought that the children studied were a highly selected group. Furthermore, any differences detected between the cases and controls were rendered more meaningful by this theoretical shortcoming of the study design.

The out of hours visiting rate in this study was very similar to that in a recent study by Pitts and Whitby,[9] who reported a rate of 152/1000 patients. Their study took place in a semiurban-semirural practice, however, and 30% of their visits were to the casualty department of the local cottage hospital. These authors thought that they had a high out of hours visiting rate and that patients now expect a 24 hour general medical service, but we would suggest that perhaps only a minority of patients demand this.

Our principal finding was that a small proportion of children generate a high proportion of the total out of hours visits to children. Frequent attenders during the day are known to generate a disproportionate volume of work, and this study confirmed the same effect at night.[10] Few of the problems for which the study children were visited out of hours required more than simple advice or reassurance. This pattern is similar to the proportion of that found by Walker[3]: our proportion of visits to children aged under 10 (20.3%) was similar to that reported by him, and the morbidity pattern was comparable (upper respiratory tract infections followed by diarrhoea and vomiting were the most common conditions). Walker also found that only 5.1% of visits resulted in admission to hospital.

It has been suggested that the number of patients who seek medical advice out of hours is underestimated because more patients attend accident and emergency departments than contact their general practitioner.[11] Our findings do not support this claim. On the contrary, few children in either the study or control group attended the accident and emergency department during the study year. Frequent attenders are also known to have more doctor defined physical illness than the rest of the community,[10] and Campion and Gabriel found that serious health problems were the most important determinants of consulting patterns in their study of illness behaviour in mothers with young children.[5] However, we found no difference in the number of doctor defined serious health problems between the study children and the controls.

Like others, we found a relation between social and economic status and frequent consultation.[5, 10] Before the study began we suspected that single mothers would be more likely to call a doctor out of hours, and our findings seem to support this. The absence of another adult in the household to consult about a child's illness may make the mother more likely to seek a second opinion from another source. For this reason we asked about the presence of close relatives nearby, but the practice is in a relatively closeknit community, so a similarly large proportion of the study group and controls had close relatives within about three miles of their home. Nevertheless, the option of discussing the problem with relatives, friends, or the chemist was infrequently chosen. The highest level of education reached was important.[5]

Other factors that are important as social indicators might also be important in their own right. For example, owning a car may mean that ill children can be brought more easily to the surgery during the day. Although the presence of a telephone in the home might be expected to be positively associated with out of hours calls, this was not the case and mothers who are worried about their children out of hours seem able to gain access to a telephone.

Although both groups of mothers assessed the severity of the vignettes similarly, the mothers of the study children were significantly more likely to consult a doctor. This was similar to their response to real-life childhood illness as these mothers presented their children more often for a medical opinion both during routine hours and out of hours. In their study Campion and Gabriel were more concerned about underconsulting due to lack of information or lack of concern by mothers than about overconsulting for trivia. In our study, however, only a small minority of the visits were for serious problems, few of the patients referred themselves to accident and emergency departments, and during the study year no children in the practice died or suffered serious illness as a result of delay in seeking medical help. Therefore underconsulting is unlikely to be an important problem at night.

Despite our findings we do not advocate that general practitioners should refuse to visit children who are seen frequently out of hours. If, however, it were possible to educate the mothers of these children to manage episodes of minor childhood illness themselves then they might become more confident about doing so. To this end the out of hours visit itself could be used as an opportunity to educate the mother, particularly if the problem is trivial. Tact is, however, required, as this study, suggests that mothers do not contact the doctor out of hours unless they feel that the situation is serious. Bollam *et al.* believed that patients would benefit from some indication of what the practice considered to be appropriate use of the out of hours service.[12] However, Usherwood has shown that written advice about particular childhood symptoms may increase out of hours visits.[13] Other opportunities for advising mothers include paediatric assessment clinics, immunisation clinics, and routine surgery attendances. The health visitor is a valuable member of the team who is underused as a source of advice on the management of childhood illness[14] — a finding confirmed by this study.

ACKNOWLEDGEMENTS

We thank Peter Campion for allowing us to use the vignette instrument; the Department of General Practice, University of Glasgow, for advice; and doctors, staff and patients who participated in the study. This study was supported by

114 General practice

a grant from the Scientific Foundation Board of the Royal College of General Practitioners. These findings were reported at the Scottish University Departments of General Practice annual scientific conference on 25 January 1991.

REFERENCES

1. Pitts, J. Hours of work and fatigue in doctors. *J R Coll Gen Pract* 1988; **38**: 2-3.
2. Tulloch, A. J. 'Out of hours' calls in an Oxfordshire practice. *Practitioner* 1994; **228**: 663-6.
3. Walker, R. D. Study of out-of-hours visits to children. *J R Coll Gen Pract* 1985; **35**: 427-8.
4. Bhopal, J. S., Bhopal, R. S. Perceived versus actual consultation patterns in an inner city practice. *J R Coll Gen Pract* 1989; **39**: 156-7.
5. Campion, P. D., Gabriel, J. Illness behaviour in mothers with young children. *Soc Sci Med* 1985; **20**: 325-30.
6. Minitab Inc. *Minitab statistical software*. Dayton: Mazer, 1986.
7. Gardner, M. J., Altman, D. G. *Statistics with confidence*. London: BMJ, 1989.
8. Machin, D., Campbell, M. J. *Statistical tables for the design of clinical trials*. Oxford: Blackwell Scientific, 1987.
9. Pitts, J., Whitby, M. Out of hours workload of a suburban general practice: deprivation or expectation. *BMJ* 1990; **300**: 1113-5.
10. Westhead, J. N. Frequent attenders in general practice: medical, psychological and social characteristics. *J R Coll Gen Pract* 1985; **35**: 337-40.
11. William, D. J. Patient's assessment of out of hours care. *BMJ* 1988; **296**: 119.
12. Bollam, M. J., McCarthy, M., Modell M. Patients' assessment of out of hours and in general practice. *BMJ* 1988; **296**: 829-32.
13. Usherwood, T. P. Development and randomised controlled trial of a booklet, advice for patients. *Br J Gen Pract* 1991; **41**: 58-62.
14. Duncan, J. K., Taylor, R. J., Fordyce, I. E. Factors associated with variation in the consultation rates of children aged under 5 year. *J R Coll Gen Pract* 1987; **37**: 251-4.

8 Multi-practice research: a cohort study

Andrew Haines

INTRODUCTION

Concerns about causation of disease are perennial amongst both doctors and their patients. Stories about possible causative factors and disease appear frequently in the media and excite considerable public interest. Issues such as the relationship between food additives and hyperactivity and between BSE and Jacob–Kreuzfeldt disease are recent examples. In order to be able to advise patients about risks to health and to make judgements about when to change their clinical practice, doctors need to be aware of how to evaluate claims of causality. Many of these claims rest upon epidemiological data which seek to identify associations between a particular factor and the occurrence of disease. Diseases are rarely monocausal but often result from the interaction of a number of factors, although one may predominate in a given individual.

TYPES OF STUDY

There are a number of types of study design which can be used in investigating causality; theoretically, it would be possible to study causation by undertaking a randomized controlled trial (RCT), exposing one group to the possible agent of interest and avoiding exposure to the control group. In this way, there is a high probability of obtaining comparable groups which do not differ significantly in potential confounding variables which might make it difficult to interpret any differences between the groups. In practice, RCT's are rarely used to study causation. If one were interested, for instance in the possible relationship between alcohol and stroke, it would be necessary to ensure that any apparent relationship was not due to the high prevalence of cigarette smoking amongst those who drink heavily. It is clearly not possible to randomize people to drink or not to drink, and therefore it is necessary to use another type of design. A cohort or prospective study is one in which a group of individuals is identified, some of whom have been exposed to the possible causative agent. The group is then followed up, usually over a considerable time period, depending on the question to be answered and the relative risk of disease in the two groups is measured. This design can often give useful information. However, the disadvantages are clear: such studies may be expensive and long-lasting because of the possibility of delayed effects in many cases. Therefore, case-control designs are sometimes used which compare a group of individuals with the condition of interest and a control group, recording a number of variables, including exposure to the potential causative factor (s) and potential confounding variables. Case-control studies may be difficult to interpret because they are more open to bias than cohort studies or RCT's. For instance the interviewer who interviews cases and controls is unlikely to be completely 'blind' to the diagnosis. Subtle biases may therefore 'creep into' the interview. Patients may have perceptions about the causation of their condition and may remember their exposure to an agent or agents more vividly than a similarly exposed control. In general, the most important deficiency of a case-control study is that information about exposure to the agent which is being investigated is necessarily obtained retrospectively, and is seriously subject to problems of recall or inadequate former recording. The other major difficulty is in finding a control population which is truly representative of the general population who have not become 'cases'. Nevertheless, these problems are very well understood, and there are many excellent case-control studies which are relatively free of these criticisms.

Some papers are published that merely describe a group of cases of a particular condition stating that an apparently high percentage of them have been exposed to a particular agent or share a certain characteristic without giving any specific data from comparison groups. If the association is strong, descriptive studies may give clues which lead on to more rigorous studies, but considered in isolation they can be misleading.

ADVANTAGES OF GENERAL PRACTICE BASED RESEARCH

The near-universal registration of patients with a GP in the UK means that studies based in general practice can survey a defined population rather than merely a group who are currently attending for care. It also minimizes potential selection biases that might operate, for example if women attending family planning clinics were to be the only group studied. They might differ in a number of ways from women not attending such clinics. For the purposes of observational studies, an even more important characteristic of the UK system is the referral system, or the gatekeeper role, as it is often called, which ensures that the general practitioner holds a comprehensive record of health service usage and diseases treated, throughout the hospital service as well as in primary care. General practitioners participating in research are a self-selected group, and it would be preferable to study randomly selected patients of randomly selected GPs. This is rarely possible, but if the intention is to study patients rather than their doctors, the use of volunteer practices may not be a serious weakness.

THE RCGP ORAL CONTRACEPTION STUDY

A well-known and influential study which looked at the potential effects of oral contraceptives on cardiovascular disease provides one example of a cohort study for critical analysis (RCGP 1981). This work is of great importance because it demonstrated that general practice was not only a suitable site for research, but that the research in question could address important questions which in this case could not have been answered adequately by hospital studies. Although this commentary focuses on data on cardiovascular disease presented in 1981, the study monitored the total range of morbidity and mortality. A more detailed description of study design, analytical methods and potential sources of bias were given in an earlier publication (RCGP 1974) and an earlier report of results in 1977 (RCGP 1977).

ASSEMBLING THE COHORT

The RCGP oral contraception study started with the advantage that participants were recruited in 1968–69 from a large number of practices (1400 general practitioners throughout the United Kingdom). However, despite the attempt to recruit a representative population, the pill users tended to be older than pill users in the general population, and the results showed that women in the study tended to be healthier and have lower death rates than the general population. This may have been due to the practices from which they were recruited being unrepresentative of the general population or because recruitment was restricted to married women (this was done because of the difficulty of identifying unmarried women who were sexually active for the control group). It is also quite commonplace to find that participants in studies have a more favourable outcome than those who refuse to participate, thus a high response rate is important. This may be partly because those at highest risk of death have priorities other than participating in research or are excluded by the study protocol, and perhaps also because being in a study results in a higher standard of medical care!

It is necessary to know the inclusion and exclusion criteria and they should be specified by the investigators in advance. Any study which excludes large numbers of individuals for whatever reasons is more difficult to interpret, since calculations of risk may be biased. For instance if women had been screened for cardiovascular risk factors such as plasma cholesterol and those with high levels had been excluded from the study, this could have resulted in an underestimate of the risk for the general population if the adverse consequences of the pill were seen particularly in those at highest risk of cardiovascular disease. In the case of the RCGP study, no such selection was undertaken, and therefore we can have reasonable confidence that the results might be generalizable. It can also be important to know how the cohort was recruited. If general practitioners were asked to recall the names of a number of women who are taking the contraceptive pill, they may have recalled the names of patients who were not typical of the population of women using oral contraception; they might have been patients who consulted more frequently and had higher levels of morbidity. Now that computerized registers are widely available, it would have been possible to take a random sample of women of a suitable age from practice age–sex registers. In the days when the study was undertaken, it was more laborious. Thus GPs were asked to recruit women taking oral contraceptives for the study when they consulted. In order to limit the workload of participating doctors, the first two women consulting for oral contraception were recruited each calender month. Controls were selected from records, working alphabetically from each pill-taker and taking the first married woman born within three years either side of the case but who had not been given oral contraception. This was checked by direct questioning (RCGP 1974).

In a cohort study, it is necessary to compare the baseline characteristics of the groups to be studied to determine their comparability. The controls were older by 0.5 year on average than the takers (RCGP 1974). Although statistically significant because of the large numbers, the difference is not likely to materially affect results and was due to the relative shortage of young married women not using the pill. Social class differences were minimal, but the takers had a higher parity than controls and takers were more likely to be cigarette smokers and to smoke heavily. The paper describing this observation was the first publication from the study (Kay et al. 1969). These differences were taken into account in the subsequent statistical analyses.

The quality of data collected during a study can be maintained in a number of ways. The use of standard data collection forms which have been piloted to ensure intelligibility and acceptability is, of course, mandatory. Coding and data entry should be undertaken and checked by a small number of trained operators. Feedback to participating doctors may help keep up interest during the course of a long study.

A study must have adequate statistical power to detect with a high degree of probability differences between the exposed and unexposed groups which are clinically significant. In writing up papers, it is helpful if investigators specify the magnitude of the differences they were hoping to detect. This informs the reader whether negative results might be due to inadequate sample size, that is inadequate 'power'. This study was designed to determine the frequency of a range of potential problems and to reveal previously unsuspected relationships with disease. Thus it was not possible to make direct calculations of the minimum populations required for study. The researchers considered four conditions which might be affected by the pill and calculated the numbers required to demonstrate a doubling of incidence at a 5 per cent level of significance (RCGP 1974). Over 23 000 current takers of oral contraceptives and a similar number of controls were recruited over a fourteeen month period. Although this seems a very large group to study, the event rates were low, particularly in women under 35 years of age. Because of the low incidence of events, estimates of excess death rates are necessarily imprecise and have wide confidence limits, particularly at younger ages.

CATEGORIZATION OF EXPOSURE

Ideally, from a scientific point of view, it would be preferable if individuals did not switch from one group to another — in this case, from taking to not taking the contraceptive pill and vice versa. Clearly in this case, exposure is necessarily determined to a great degree by the wishes of the patient and it is therefore impossible to ensure. It is important to set up rules in advance about how participants who change category will be treated. Clearly, if former users are classified with current users, this could considerably dilute any potential adverse effects of the pill if they were short-lived. In the RCGP study, for each month of observation, participants are classified either as current users or former users of oral contraceptives or as controls. For some analyses, current and former users were added together as ever users, but for others they were separated. Consumption of the contraceptive pill is relatively easy to study because its availability in the UK is restricted by the need to prescribe. In the case of many other variables such as diet, exposure to atmospheric pollutants, etc. measurement of exposure may be fraught with difficulty and can result in substantial underestimates of effect.

OUTCOME MEASURES

Aetiological studies commonly have a number of outcome measures, the most important of which is mortality. Death is likely to be well recorded. In a young age group such as the one studied in the RCGP study, the cause of death is likely to be relatively accurate. Even so, it is important that those who classify the cause of death for study purposes are blind to whether the individual patient has been exposed to the variable of interest — in this case, the contraceptive pill. This prevents unconscious biases in the assignment of diagnosis. Many studies use standardized criteria for the diagnosis of deaths and events, for example the WHO criteria for diagnosis in myocardial infarction (WHO 1976). This ensures comparability between studies and wide acceptance of the results. In the RCGP study, the underlying cause of death was coded by one investigator and checked by one of two others using the eighth revision of the International Classification of Diseases (ICD) (1967). In the case of morbidity information, the data is often softer but in many studies, death is too rare an event or too extreme an outcome to be the primary variable of interest. Increasingly, instruments are being developed to measure quality of life and to assess the cost-effectiveness of different interventions, and these are data which are required for planning the implementation of specific treatments or management strategies.

FOLLOW-UP

It is important that follow-up of participants is as complete as possible. If large numbers of individuals are lost from the cohort over the period of follow-up, the results may be seriously affected. It is clear that significant events could be missed and that there is a potential for bias due to loss of an unrepresentative group. For example if there had been disproportionate loss of older women who smoked and took the pill and who were therefore at relatively high risk of vascular events, this could have resulted in an underestimate of the risks due to the pill.

In the UK, notification of death certification is likely to be high, although certificates may take sometime to reach the investigators. Individuals can be tagged in the National Health Service Central Registry, for which a fee is paid. When the individual dies, notification is sent to the study team. In the case of the RCGP study, general practitioners' reports were used for analyses of mortality. The investigators demonstrated (Wingrave et al. 1981) that these were as valid as the use of death certificates, but they obviously depend on the investigators being in contact with the general practitioner looking after the patient at the time of their death. The completeness of follow-up of the cohort is not clear from the 1981 RCGP paper. The 1981 report had been preceeded by a report on mortality of women in the study published in 1977, so it would have been useful to know how many had been lost to follow-up in the intervening years, although in this case, it is unlikely to affect the interpretation of the results.

STATISTICAL ANALYSIS

An important comparison for aetiological purposes is between the rates of events — in this case mortality — in those who are exposed and those not exposed, as this tells us the number of times more likely it is for a woman taking the pill to die of a given cause than one who is not taking the pill during the period of follow-up. However, from a clinical perspective, it is also important to know the excess risk, because this helps to put the risk in perspective. For instance Table 2 in the paper shows us that ever users of the pill are 4.2 times more likely to die from circulatory diseases than never users. This is obviously worrying, but the excess risk is 22.7 deaths per hundred thousand woman years (i.e. 100 000 women followed for one year or 10 000 women followed for 10 years). Thus, the risk to the individual is relatively small despite the substantially elevated relative risk because deaths due to cardiovascular diseases are rare in young women. Popular reports of studies often focus on the relative risk and not the excess risk, which may give a misleading impression of the danger to the individual of exposure to a given 'hazard'.

The figures given are of course estimates, and, like any other quantitative measure, are open to sampling error. Thus, it is important to look at the 95 per cent confidence intervals. These are sometimes called confidence limits, as in the RCGP paper, but as this implies absolute boundaries, the term confidence intervals may be preferable. A 95 per cent confidence interval, is the interval within which we can have 95 per cent confidence that the 'true' value lies (i.e. the value in the whole population from which the study sample is recruited). The distance apart of the confidence intervals indicates how much reliance we can place on the estimate. For example in the case of mortality for all circulatory diseases in the study, the relative risk related to pill exposure is 4.2. 95 per cent confidence intervals lie between 2.3 and 7.7. Since the confidence intervals do not embrace unity, there is a high probability that this difference is not due to chance alone.

Two-tailed tests were used in the study, as they are in most statistical analyses. These imply that the differences could have been in either direction. If there was a very strong a priori hypothesis that the difference could only be in one direction, some investigators would use a one-tailed test, which gives a higher level of significance for a given difference. Most, however, would take a more conservative view, as has been done in this paper. The results presented have also been standardized for age, parity, cigarette consumption, and social class at entry into the study. This is important because differences in social class or cigarette smoking between contraceptive users and the control group might increase or weaken the relative risk which is attributed to the use of the contraceptive pill. Mortality rates were calculated by dividing the number of deaths in each pill usage group by the calendar months of follow-up. This makes it possible to take into account the different number of women in each group and the length of time for which they were observed.

INTERPRETATION OF RESULTS

It is important to know whether any individuals who were entered into the cohort were subsequently excluded from it for any reason, as this could create bias. In the RCGP study, only 33 deaths were excluded for the 1981 report (compared with 249 which were reported), because they died from conditions diagnosed before recruitment into the study. The numbers were similar in the pill user and control group (17 and 16 respectively) and the investigators decided to exclude them on the grounds that the disease from which they were suffering may have influenced contraceptive choice. The numbers excluded were small and it is unlikely that they had any perceptible effect on the estimates of risk.

Clinicians would wish to know whether there are any groups within the study population which were particularly at risk of complications from the pill so that in future they could be counselled to avoid this method of contraception. The investigators therefore looked within the group to see whether they could define sub-groups at particularly high risk. The obvious factors of interest were age and smoking, and they showed a much greater excess annual death rate in women aged 35 to 44, and an even higher rate in those aged 45 and above, compared with younger women. Because of the rarity of death in women under 35, the estimates of excess annual death rates had wide confidence intervals and could have been a chance finding.

They also looked at specific causes of death and demonstrated that two, namely ischaemic heart disease and subarachnoid haemorrhage, were significantly more frequent in ever users than in controls. There was a high level of diagnostic certainty in the cases of subarachnoid haemorrhage, with 75 per cent being confirmed at necropsy. Six of the deaths from subarachnoid haemorrhage were amongst current pill users (the number 5 is given in the text – 6 cases are listed in the table). Eleven were amongst former users and only 3 amongst controls. The authors discussed the possible significance of the association, and referenced three studies which had found a similar possible relationship and one that did not.

The study also looked at the important question of whether complications were related to duration of pill use. In contrast to an earlier paper (RCGP 1977), they found no association between increasing risk of death due to vascular disease with increasing duration. This may be because in the first paper, the investigators were only able to divide women into two groups of users: continuous pill usage of less than 5 years or 5 years and longer. In the 1981 paper, they were able to look at more finely graded durations of use and they also standardized for smoking and parity as well as age.

The raised excess risk of former users is clearly potentially important, but there are problems in interpretation. An excess risk in former users could be interpreted as evidence for a long-term effect of the pill, but it could also be due to women who are ill being withdrawn from the pill. In order to differentiate between these two explanations, it is important to look in detail at the individual cases. The authors point out that for non-rheumatic heart disease and hypertension, four of the ten deaths among former users were amongst women who showed symptoms of their condition that was ultimately to kill them when still taking the pill. If these are classified as current users, the excess risk in former users is no longer substantially increased. This also indicates that even in a large study such as this one, differences in the classification of even a small number of cases can affect conclusions substantially.

What other explanations could be put forward for the relationship of the pill and cardiovascular disease besides a causal one? Could there be substantial bias in the way in which deaths were recorded and notified by general practitioners. If this is so, one might expect to see an increase in non-cardiovascular mortality in the pill group, but this is not found. Could it be due to differences in characteristics between pill users and non-users? Clearly, since the study was not a randomised trial, we cannot entirely exclude this possibility, although it seems remote as the investigators adjusted as far as they could for important potential confounding factors. One proviso is that smoking was only recorded at entry to the study and thus women could have changed their smoking status during the course of the study. This is more likely to diminish slightly the apparent adverse effect of smoking, because of the mis-classification of former smokers as current smokers, rather than to substantially influence the excess risk of pill users. Confounding is derived from the Latin *confundere*, to mix together. It describes a situation in which the effects of two processes are not separated. There is a distortion of the apparent effect of an exposure on risk brought about by the association with other factors that can independently influence the outcome. A confounding variable can only be controlled by statistical means in order to obtain an undistorted estimate of the effect of the study factor on risk if it can be measured (Last 1988). Confounding is a ubiquitous problem in research and investigators must always to be on the alert to exclude confounding as a possible explanation for apparently causal associations.

CRITERIA FOR JUDGING WHETHER A RELATIONSHIP IS CAUSAL

Some years ago, Sir Austin Bradford Hill, the renowned pioneer of medical statistics, described a number of criteria for deciding whether an association was likely to be causal (Bradford Hill 1965). The Department of Clinical Epidemiology and Biostatistics at McMaster University Health Sciences

Centre in Canada have a long interest in improving health professionals' ability to read medical literature critically. In a series of papers in the Canadian Medical Association Journal in 1981 (Department of Epidemiology and Biostatistics, McMaster University, Health Sciences Center 1981*a*/*b*), and recently updated in the Journal of the American Medical Association (Guyatt *et al.* 1993, 1994) they articulated methods for reading clinical journals. They slightly adapted Bradford Hill's criteria and proposed that having decided whether the basic methods used to study causation were strong (with randomized trials being the strongest and case series being the weakest), there were a number of 'diagnostic' tests that could be applied to decide if causation seemed a likely explanation for a given association. These are summarized in Table 8.1. Applying these criteria to the data from the RCGP study, it is clear that point 1 — 'Is there evidence from true experiments in humans?'—is fulfilled as well as it can be in the absence of a randomized trial. The data from the study also suggest a strong association when potential confounding factors are taken into account. The results are consistent with other publications on this topic. Since this is the largest prospective study on health effects of the pill, these results can be given more weight than those from smaller studies and those with a retrospective case-control design (Shapiro *et al.* 1979; Vessey *et al.* 1977, 1981).

The appropriate temporal relationship clearly depends on the mechanism by which oral contraceptives cause an increase in cardiovascular mortality. The fact that there is no evidence from this study that duration of use increases risk amongst current users, could be construed as evidence against a major effect on long-term conditions such as atheroma and rather more in favour of other mechanisms such as an impact on blood pressure and perhaps haemostatic variables (Meade *et al.* 1977). However, even with these large numbers and a considerable period of follow-up, it could still be argued that longer periods of follow-up are necessary to detect a relationship between duration of use and mortality if there was an additional long-term adverse impact of pill taking.

Table 8.1 *Tests for causation (Adapted from The environment and disease: association or causation? Bradford Hill 1965)*

---

1. Is there evidence from true experiments in humans?
2. Is the association strong?
3. Is the association consistent from study to study?
4. Is the temporal relationship appropriate?
5. Is there a dose-response gradient?
6. Does the association make epidemiological sense?
7. Does the association make biological sense?
8. Is the association specific?
9. Is the association analogous to a previously proven causal association?

This study does not address the issue of whether or not there is a dose response gradient for the oestrogen component of the pill because there was insufficient data. The majority of participants were taking what is now considered to be a relatively high dose pill with $50\,\mu$g oestrogen. However, a later publication examined the association between dose of progestogen and arterial disease (Kay 1982). It was found that reports of cardiovascular diseases and total arterial diseases showed a significant trend with increased dose of norethisterone acetate. Total arterial diseases were also reported more frequently in women taking contraceptives with $250\,\mu$g rather than $150\,\mu$g of levonorgestrel. The association described between the pill and cardiovascular mortality makes epidemiological sense; is compatible with other studies and the study is appropriately designed. Potential biases and confounding variables have been considered in the analysis.

The association makes biological sense in that the pill is known to affect cardiovascular risk factors, including blood pressure, lipids, and coagulation factors. It appears specific in as much as it applies to diseases of the circulatory system and not to other categories of disease. The aetiology of ischaemic heart disease is likely to be different from that of subarachnoid haemorrhage, although high blood pressure is a risk factor in both cases.

The final and weakest yardstick for the assessment of possible causal relationships is that of analogy to a previously proven caused relationship. Men given large doses of oestrogen for cancer of the prostate have raised cardiovascular mortality (Johansson *et al.* 1991). Post-menopausal women given oestrogen replacement appear to have a lower cardiovascular mortality (Stampfer and Colditz 1991). In the latter case, the dose of oestrogen is much lower than in the contraceptive pill and oestrogens may partially counteract the increased risk of cardiovascular disease which occurs post-menopausally (Meade *et al.* 1983). However, progestogens may have an opposing effect on cardiovascular risk in post-menopausal women (Goldman and Tosteson 1991) and studies are currently underway to examine the relative effects of opposed and unopposed oestrogens on cardiovascular disease and morbidity and mortality from all causes in post-menopausal women.

CONCLUSIONS

At first sight, studies of this kind might seem expensive because of their large size and long period of follow-up. In fact, in relation to the importance of the information, they represent good value for money. The budget for the RCGP study was approximately £115 000 per annum at 1989 prices (Kay personal communication). This is likely to be much less than a similar study would have cost in many other countries which lack the infrastructure for general practice research. In such circumstances, it may not have even been feasible.

There are a number of ways in which general practitioners can become involved in multicentre studies. In addition to studies mounted by the RCGP, the Medical Research Council General Practice Research Framework (based in the Department of Environmental and Preventive Medicine at St Bartholomew's Medical College London) coordinates a range of major studies in general practice (see Chapter 9; MRC Working Party 1992). The Health Behaviour Unit at the Institute of Psychiatry has undertaken a number of studies of smoking cessation in general practice (e.g. Russell *et al.* 1988). There are also increasing numbers of networks of general practices based either on diseases or regions. The organization and management of large studies involving large numbers of practitioners can be difficult. An attempt to involve large numbers of Italian general practitioners in a trial of treatment of isolated systolic hypertension ended in failure because only just over 10 per cent of general practitioners who said they would participate actually contributed patients (Tognoni *et al.* 1991). Not all investigations require large studies; it is possible to undertake useful research in a single practice if the condition or problem being studied is common and the outcome of interest is frequently observed. Sample size calculations will indicate whether a given size of study is likely to provide sufficient power to demonstrate clinically important differences (Pocock 1989).

This study demonstrates the potential of general practice for collaborative research on issues of major clinical importance. It is a landmark study which made a major contribution to understanding the impact of oral contraceptives on mortality and it led to important clinical recommendations about when consideration should be given to stopping use of the pill—particularly in older women who also smoke. It is a significant contribution to world literature on this subject and a standard against which many other less well designed and executed studies can be compared.

REFERENCES

Bradford Hill, A. (1965). The environment and disease: association or causation? *Proceedings of the Royal Society of Medicine*, **58**, 295–300.

Dept of Clinical Epidemiology and Biostatistics, McMaster University Health Sciences Center (1981*a*). How to read clinical journals: IV. To determine etiology or causation. *Canadian Medical Association Journal*, **24**, 985–90.

Dept of Clinical Epidemiology and Biostatistics, McMaster University Health Sciences Center (1981*b*). How to read clinical journals: V. To distinguish useful from useless or even harmful therapy. *Canadian Medical Association Journal*, **24**, 1156–62.

Goldman, L. and Tosteson, A. N. A. (1991). Uncertainty about postmenopausal estrogen. *The New England Journal of Medicine*, **325**, 800–2.

Guyatt, G. H., Sackett, D. L., and Cook, D. J. (1993). Users' guides to the medical literature. II. How to use an article about therapy or prevention A. Are the results of the study valid? *Journal of the American Medical Association*, 270, 2598-2601.

Guyatt, G. H., Sackett, D. L., and Cook, D. J. (1994). Users' guides to the medical literature II. How to use an article about therapy or prevention B. What were the results and will they help me in caring for my patients? *Journal of the American Medical Association*, 271, 59-63.

International Classification of Diseases (1967). *Manual of the international statistical classification of diseases.* World Health Organisation, Geneva.

Johansson, J. E., Andersson, S. O., Holmberg, L., and Bergstrom, R. (1991). Primary orchiectomy versus estrogen therapy in advanced prostatic cancer – a randomized study: results after 7 to 10 years of follow-up. *Journal of Urology*, 3, 519-22.

Kay, C. R. (1982). Progestogens and arterial disease – evidence from the Royal College of General Practitioners' Study. *American Journal of Obstetrics and Gynecology*, 6, 762-5.

Kay, C. R., Smith, A., and Richards, B. (1969). Smoking habits of oral contraceptive users. *Lancet*, ii, 1228-9.

Last, J. M. (ed.) (1988). *A dictionary of epidemiology.* International Epidemiological Association. Oxford University Press, New York.

Meade, T. W., Chakrabarti, R., Haines, A. P., Howartt, D., North, W. R. S., and Stirling, Y. (1977). Haemostatic lipid and blood pressure profiles of women on oral contraceptives containing 50 mg or 30 mg oestrogen. *Lancet*, ii, 948-51.

Meade, T. W., Haines, A. P., Imeson, J. D., Stirling, Y., and Thompson, S. G. (1983). Menopausal status and haemostatic variables. *Lancet*, i, 22-4.

MRC Working Party (1992). Medical Research Council trial of treatment of hypertension in older adults: principal results. *British Medical Journal*, 304, 405-11.

Pocock, S. J. (1989). *Clinical trials, a practical approach.* John Wiley & Sons, Chichester.

Royal College of General Practitioners (1974). *Oral contraceptives and health.* Pitman Medical, London.

RCGP Oral Contraception Study (1977). Mortality among oral contraceptive users. *Lancet*, ii, 727-31.

RCGP Oral Contraception Study (1981). Further analyses of mortality in oral contraceptive users. Occasional Survey. *Lancet*, i, 541-6.

Russell, M. A. H., Stapleton, J. A., Hajek, P., Jackson, P. H., and Belcher, M. (1988). District programme to reduce smoking: Can sustained intervention by general practitioners affect prevalence? *Journal of Epidemiology and Community Health*, 42, 111-5.

Shapiro, S., Sloane, D., Rosenberg, L., Kaufman, D. W., Stolley, P. D., and Mieltinen, O. S. (1979). Oral contraceptive use in relation to myocardial infarction. *Lancet*, i, 743-7.

Stampfer, M. J. and Colditz, G. A. (1991). Estrogen replacement therapy and coronary heart disease: a quantitative assessment of the epidemiologic evidence. *Preventive Medicine*, 20, 47-63.

Tognoni, G., Alli, C., Avanzini, F., Betteli, G, Colombo, F., and Curso, R., *et al.* (1991). Randomised clinical trials in general practice: lessons from a failure. *British Medical Journal*, **303**, 969-72.

Vessey, M. P., McPherson, K., and Johnson, B. (1977). Mortality among women participating in the Oxford/Family Planning Association Contraceptive Study. *Lancet*, **ii**, 731-3.

Vessey, M. P., McPherson, K., and Yeates, D. (1981). Mortality in oral contraceptive users. *Lancet*, **i**, 549.

Wingrave, S. J., Beral, V., Adelstein, A. M., and Kay, C. R. (1981). Comparison of cause of death coding on death certificates with coding in the Royal College of General Practitioners Oral Contraception Study. *Journal of Epidemiology and Community Health*, **35**, 51-8.

World Health Organisation Regional Office for Europe (1976). Myocardial infarction community registers. *Public Health in Europe*, **5**. WHO, Copenhagen.

Further analyses of mortality in oral contraceptive users

Royal College of General Practitioners' Oral Contraception Study*
The Lancet, 541-6 (1981)

Abstract

An analysis has been made of the 249 deaths that have been reported during the course of the Royal College of General Practitioners' prospective study of the health associations of oral contraception. As previously, women who had used the pill were reported to have a 40% higher death rate than women who had never taken oral contraceptives. Virtually all the excess mortality was due to diseases of the circulatory system. Women who had used the pill had a relative risk of 4.3 for deaths attributed to vascular diseases. Most of these were from ischaemic heart disease (relative risk 4.1) and subarachnoid haemorrhage (relative risk 4.0). The larger number of cases now reported has permitted greater discrimination than was possible earlier between subjects with high and low risks. There were few deaths under 35 years of age and the excess annual death rates of 1 per 77 000 women in non-smokers and 1 per 10 000 women in smokers have wide confidence limits, and could have arisen by chance. For women aged 35-44 years the excess annual death rate was 1 per 6700 women in non-smokers and 1 per 2000 women in smokers; at 45 years and above the risks were 1 in 2500 and 1 in 500, respectively. There is now no evidence that the risk is associated with duration of oral contraceptive use. The demonstrated positive association of risk with parity in pill users needs confirmation. Former-users were noted to have an increased risk of death from vascular disease, but it was not possible to determine whether this represents a residual effect of oral contraceptives on the vascular system.

INTRODUCTION

The Oral Contraception Study of the Royal College of General Practitioners is the largest prospective survey of the health effects of the pill. The last report on the mortality of women in this continuing study, published in 1977,[1] indicated that the relative risk of death from a circulatory disease associated with ever-use of oral contraceptives was 4.7. This relative risk increased with age, cigarette smoking, and the duration of pill use. A large part of the excess mortality from circulatory disorders in women who had used the pill was due to subarachnoid haemorrhage, which had not been previously associated with pill usage. The number of deaths now available for analysis has more than doubled since the earlier report. The previous report included only women whose pill use had been continuous since entry into the study. Women with intermittent oral contraceptive use are now included in the analyses.

The larger numbers allow us to study some unresolved issues, such as the relation between cigarette smoking and oral contraceptives in women of various ages, and the risk of death from circulatory diseases in former-users of oral

*Principal authors: Dr PETER M. LAYDE and Dr VALERIE BERAL, Department of Medical Statistics and Epidemiology, London School of Hygiene and Tropical Medicine, London WC1E 7HT; and Dr CLIFFORD R. KAY (Director, R.C.G.P. Oral Contraception Study), 8 Barlow Moor Road, Manchester M20 0TR.

contraceptives. In addition, we have re-examined our earlier findings on the effect of duration of pill use on the risk of circulatory mortality, and on the risk of subarachnoid haemorrhage in ever-users of the pill.

METHODS

The design of the study has been described in detail previously.[2] Briefly, during a 14-month period in 1968 and 1969, over 23 000 women taking oral contraceptives and an equal number of controls who had never taken the pill were recruited by 1400 general practitioners throughout the United Kingdom. Twice a year the general practitioners report details on contraceptive use and morbidity and mortality for each of their study subjects. The doctors report the cause of death in the same form as is required for statutory death certification. We have used the general practitioners' reports for our analyses of mortality and we have demonstrated elsewhere[3] that these are as valid for our purpose as the use of the death certificates. The underlying cause of death was assigned by C.R.K., coded according to the 8th Revision of the International Classification of Diseases (ICD),[4] and checked by V.B. or P.M.L.

For each month of observation, women are classified either as current-users or former users of oral contraceptives, or as controls. For most analyses, current-users, and former-users are grouped together as ever-users. Controls who later became pill users are included in the appropriate user group from the time of the change. Women who have resumed oral contraceptive use are included in the current-user group from the time of resumption. This is in contrast to the previous mortality report where they were excluded. In this paper, current-users are all women who were currently using the pill; their use may have been continuous or they may have resumed pill use after stopping it on one or more occasions. Similarly, former-users are all women who had used the pill previously but were not currently doing so.

Mortality rates are calculated by dividing the numbers of deaths in each pill usage group by the calendar months of observation for women in that group. They are expressed as rates per 100 000 woman-years. Unless otherwise specified, mortality rates are adjusted by indirect standardisation for age and parity at death, and for cigarette consumption and social class at entry into the study, the total study population being taken as the standard.[2] Because the rates for each cause of death are standardised separately, there are small discrepancies between the sum of the individual rates and the rate for total mortality.

The only deaths excluded from these analyses were of 33 women (17 controls and 16 ever-users) who died from conditions diagnosed before their recruitment into the study, since these conditions could have influenced contraceptive choice. These exclusions were made without knowledge of the contraceptive usage of the women. All periods of pregnancy and related deaths are included.

Statistical significance has been calculated by a two-tailed χ^2 test developed by Peto.[2] Two measures of the relationship between oral contraceptive use and mortality are used.[5] Relative risk is the ratio of the mortality rate in the user groups to that of controls. Excess risk is the difference in the number of deaths per 100 000 woman-years between the user groups and controls. Confidence limits for the relative risks and excess risks were estimated by Miettenen's method.[6] When the 95% confidence limits of the relative risk do not include unity, or those of the excess risk do not include zero, the difference between pill users and controls was statistically significant at the 0.05 level or lower. For factors with more than two levels, the significance of linear trends was determined by Mantel's test.[7]

This report is based on data up to December, 1979. The periods of observation total 98 997 woman-years for current-users, 84 811 for former-users, and 138 630 for controls.

RESULTS

Ever-users of oral contraceptives had a 40% higher overall mortality rate than the controls (table I). The excess risk among the ever-users was 23.3 per 100 000 woman-years. The increased overall death rate in the ever-users was almost entirely due to their excess deaths from diseases of the circulatory system. For circulatory deaths the ever-users had a relative risk of 4.2 and an excess risk of 22.7 per 100 000 woman-years.

Table II gives a more detailed analysis of deaths from diseases of the circulatory system. For the category of non-rheumatic heart disease and hypertension (ICD 400–429) the ever-users had a relative risk of 5.6. 17 out of the 24 deaths in this category were from ischaemic heart disease, where ever-users had a relative risk of 3.9. For all cerebrovascular diseases (ICD 430–438) ever-users had a relative risk of 2.9. Here, 20 out of the 34 deaths were from subarachnoid haemorrhage, where the relative risk for ever-users was 4.0; for strokes from other causes (cerebral haemorrhage, thrombosis, and embolus) the relative risk was 2.1. Amongst other vascular diseases (ICD 440–458) 5 of the 7 deaths were from pulmonary embolism; the relative risk could not be determined because there were no deaths from these causes among the controls.

When oral contraceptive use at the time of death is considered (table III) the relative risk of circulatory mortality was 4.0 for current-users and, similarly, 4.3 for former-users. In the category of non-rheumatic heart disease and hypertension, current-users had a higher relative risk than former-users — 7.3 and 4.6, respectively. But for cerebrovascular disease the reverse was found: current-users had a relative risk of 2.0, whereas former-users had a relative risk of 3.6.

The increased risk in former-users could result from the development of ultimately fatal illnesses in some women while they were current-users and which led to their stopping the pill. We therefore reviewed the records of the 32 former-users who died of circulatory causes. Of the 10 women who died of non-rheumatic heart disease and hypertension, 4 (40%) acquired the condition from which they eventually died while still using the pill (3 cases of malignant hypertension and 1 of coronary thrombosis). In addition, 1 woman (10%) became hypertensive (blood-pressure 210/115 mm Hg) while taking the pill, and as a result oral contraception was discontinued. She eventually died of a myocardial infarction. Among the 19 former-users who died of cerebrovascular disease, 1 (5%) had a stroke while still taking the pill. In addition, 10 experienced morbidity possibly associated with their eventual death — hypertension in 5, headaches in 4, and suspected deep-vein thrombosis in 1 — while still using oral contraceptives.

Table I *Mortality rates per 100 000 woman-years by oral contraceptive use*

ICD code	Cause	Standardised mortality rate (no. of deaths)		Relative risk (95% confidence limits)	Excess risk (95% confidence limits)
		Ever-users	Controls		
140–209	All cancers	30.0 (53)	31.5 (46)	1.0 (0.5, 1.9)	−1.5 (−22.3, 19.3)
390–458	All diseases of the circulatory system	29.9 (55)	7.2 (10)	4.2 (2.3, 7.7)	22.7 (13.2, 32.2)
630–678	Complications of pregnancy, childbirth, and puerperium	1.0 (2)	0.8 (1)	1.2 (0.5, 2.7)	0.2 (−0.7, 1.1)
E800–E999	Accidents and violence	18.2 (34)	12.6 (17)	1.5 (0.8, 3.0)	5.6 (−3.8, 15.0)
	All other causes	7.8 (12)	11.2 (19)	0.7 (0.3, 1.6)	−3.4 (−11.2, 4.5)
	Total	87.7 (156)	64.4 (93)	1.4 (1.1, 1.8)	23.3 (4.7, 41.9)

Table II *Mortality rate per 100 000 woman-years from various diseases of the circulatory system by oral contraceptive use*

ICD code	Cause	Standardised mortality rate (no. of deaths)		Relative risk (95% confidence limits)	Excess risk (95% confidence limits)
		Ever-users	Controls		
400–429	All non-rheumatic heart disease and hypertension	11.8 (21)	2.1 (3)	5.6 (2.0, 16.6)	9.7 (3.8, 15.6)
400	Malignant hypertension	1.7 (3)	0.0 (0)	–	1.7 (−1.5, 4.9)
410–414	Ischaemic heart disease	8.0 (14)	2.0 (3)	3.9 (1.2, 12.9)	6.0 (0.7, 11.3)
430–438	Cerebrovascular diseases	14.7 (27)	5.0 (7)	2.9 (1.3, 6.4)	9.7 (2.6, 16.8)
430	Subarachnoid haemorrhage	9.0 (17)	2.3 (3)	4.0 (1.3, 12.9)	6.7 (1.0, 12.4)
431–433	Cerebral thrombosis, haemorrhage, and embolism	5.7 (10)	2.7 (4)	2.1 (0.6, 7.9)	3.0 (−2.3, 8.3)
440–458	Other vascular diseases	3.4 (7)	0.0 (0)	–	3.4 (−0.5, 7.3)
450–453	Pulmonary embolism and thrombophlebitis	2.5 (5)	0.0 (0)	–	2.5 (−1.0, 6.0)
390–458	All circulatory diseases	29.9 (55)	7.2 (10)	4.2 (2.3, 7.7)	22.7 (13.2, 32.2)

Table III *Mortality rate per 100 000 woman-years from various diseases of the circulatory system by oral contraceptive use at time of death*

ICD code	Cause	Standardised mortality rate (no. of deaths)			Relative risk (95% confidence limits)	
		Current-users	Former-users	Controls	Current-users	Former-users
400–429	*All non-rheumatic heart disease and hypertension*	15.1 (11)	9.6 (10)	2.1 (3)	7.3 (2.0, 26.7)	4.6 (1.3, 16.5)
400	Malignant hypertension	0.0 (0)	2.5 (3)	0.0 (0)	—	—
410–414	Ischaemic heart disease	13.0 (10)	4.1 (4)	2.0 (3)	6.4 (1.7, 23.5)	2.0 (0.2, 17.6)
430–438	*Cerebrovascular diseases*	10.1 (18)	18.2 (19)	5.0 (7)	2.0 (0.6, 6.2)	3.6 (1.5, 8.5)
430	Subarachnoid haemorrhage	7.3 (6)	10.2 (11)	2.3 (3)	3.2 (0.6, 16.3)	4.5 (1.2, 16.5)
431–433	Cerebral thrombosis, haemorrhage, and embolism	2.7 (2)	8.1 (8)	2.7 (4)	1.0 (0.8, 1.2)	3.0 (0.7, 12.0)
440–458	*Other vascular diseases*	3.6 (4)	3.1 (3)	0.0 (0)	—	—
450–453	Pulmonary embolism and thrombophlebitis	2.8 (3)	2.2 (2)	0.0 (0)	—	—
390–458	*All circulatory diseases*	28.6 (23)	30.9 (32)	7.2 (10)	4.0 (1.9, 8.3)	4.3 (2.2, 8.4)

Because of the interest in the association of oral contraceptive use and subarachnoid haemorrhage, we performed a more detailed analysis of the twenty deaths from this cause. For current-users the relative risk of subarachnoid haemorrhage was 3.2, while that for former-users was 4.5 (table III). None of the 11 former-users who died of a subarachnoid haemorrhage had an initial stroke while still on the pill (table IV). 2 had hypertension and 3 had headaches while using oral contraceptives. In only 2 cases (1 hypertension and 1 headache) were these conditions stated as the reason for stopping pill use. Of the 5 current-users who died of subarachnoid haemorrhage, 1 had been diagnosed as hypertensive and 3 had had headaches before the subarachnoid haemorrhage. One current-user had a coarctation of the aorta repaired at age 23, and 1 control had mitral stenosis.

Cigarette smoking was more common in women who died of subarachnoid haemorrhage than in other women in the study: 71% (12/17) of the ever-users and 67% (2/3) of the controls who died of subarachnoid haemorrhage smoked compared with 48% of all ever-users and 40% of all controls.

The level of diagnostic certainty was high in the subarachnoid haemorrhage deaths: 75% (15/20) of the diagnoses were confirmed at necropsy and 5% (1/20) at neurosurgery, the remaining 4 (20%) being made on clinical grounds.

Table V shows the relationships of pill use, age, and smoking with circulatory mortality. The rates are specific for each age, smoking, and pill usage group, and are not standardised for parity or social class. The relative risk for ever-use of the pill increased with increasing age among both smokers and non-smokers – apart from the 15 – 24-year age group where there was only one death. For all women (smokers and non-smokers) under 35 years of age, the relative risk was 2.8 (95% confidence limits: 0.6, 12.7), while for all women 35 years and older the relative risk was 4.8 (2.4, 9.5).

The relative risk for ever-users was greater among smokers than among non-smokers for each age group. With all ages combined, the age-standardised relative risk associated with oral contraceptive use in smokers was 5.1 (2.4, 10.9) while that in non-smokers was 3.2 (1.1, 9.0). The relative risk of pill use was particularly high among older women who smoked. For women aged 45 and older, ever-users who smoked had a relative risk of 18.1 compared with their non-smoking controls.

The relationship of oral contraceptive use and parity with circulatory mortality is shown in table VI. The relative risk for ever-users rose with increasing parity. For ever-users death rates increased with increasing parity (test for trend, p < 0.001), but among controls there was no apparent association between parity and death rates due to circulatory diseases. The positive association with parity was also found for both current and former-users when they were analysed separately.

Table VII shows the relation of the total duration of current oral contraceptive use with circulatory mortality. In this table the duration of pill use is the total time the woman had used oral contraceptives, whether the use was continuous or intermittent, and includes continuous use before entry into the study. Rates in all of the duration-of-use groups were higher than the control rate. When all the groups were considered, including the controls, there was a highly significant trend (p < 0.001). However, there was no evidence of an effect of duration of use amongst the current-users (p = 0.85). The significant overall trend was due

Table IV *Summary of deaths from subarachnoid haemorrhage*

Pill usage	Age	Smoker?	Date of death	Date pill stopped	Morbidity during current pill use	Reasons for stopping pill	Confirmation of diagnosis
1. Current	35	Yes	6/69	—	None	—	Post-mortem
2. Current	35	Yes	2/75	—	Migraine	—	Post-mortem
3. Current	41	Yes	9/72	—	Hypertension	—	Post-mortem
4. Current	41	Yes	3/74	—	Tension headaches	—	Clinical
5. Current	45	No	2/73	—	None	—	Post-mortem
6. Current	39	Yes	12/69	—	Migraine	—	Post-mortem
7. Former	47	Yes	9/77	5/77	Hypertension	1. Mild hypertension 2. Amenorrhoea	Post-mortem
8. Former	43	Yes	6/77	3/69	None	Desire to conceive	Post-mortem
9. Former	37	No	6/71	2/71	None	1. Worried about long-term effects 2. Tubal sterilisation	Post-mortem
10. Former	46	No	7/71	3/70	Hypertension	Hysterectomy	Clinical
11. Former	32	No	8/72	1/71	Migraine, obesity, vestibular neuronitis	Death of husband	Post-mortem

Table IV (*Contd.*)

Pill usage	Age	Smoker?	Date of death	Date pill stopped	Morbidity during current pill use	Reasons for stopping pill	Confirmation of diagnosis
12. Former	40	No	5/79	11/72	None	1. Desire to conceive 2. Worried about long-term effects	Post-mortem
13. Former	46	Yes	4/73	7/72	°Obesity	1. Weight gain 2. Anxiety	Post-mortem
14. Former	44	Yes	1/76	6/71	Headaches	1. Headaches 2. Loss of libido	Post-mortem
15. Former	33	Yes	10/71	7/69	Aching legs, felt sick	Aching legs	Post-mortem
16. Former	53	Yes	9/77	3/71	Headaches	Menopause	Clinical
17. Former	46	Yes	1/73	11/71	None	Felt ill	Neurosurgery
18. Control	37	Yes	7/71	—	Hypertension*	—	Post-mortem
19. Control	51	Yes	3/76	—	Migraine*	—	Post-mortem
20. Control	41	No	6/76	—	—	—	Clinical

*Controls have never used the pill. Relevant morbidity reported during observation in the study is noted.

Table V *Circulatory disease mortality rates per 100 000 woman-years by age, smoking status, and oral contraceptive use*

| | Mortality rate (no. of deaths) | | Ever-users vs controls | |
| | | | | |
Age	Ever-users	Controls	Relative risk (95% confidence limits)	Excess risk (95% confidence limits)
15–25				
Non-smokers	0.0 (0)	0.0 (0)	—	0.0
Smokers	10.5 (1)	0.0 (0)	—	10.5 (−17.4, 38.4)
25–34				
Non-smokers	4.4 (2)	2.7 (1)	1.6 (0.2, 17.4)	1.7 (−6.6, 10.0)
Smokers	14.2 (6)	4.2 (1)	3.4 (0.5, 24.6)	10.0 (−6.5, 26.4)
35–44				
Non-smokers	21.5 (7)	6.4 (2)	3.3 (0.8, 14.7)	15.1 (−3.4, 33.5)
Smokers	63.4 (18)	15.2 (3)	4.2 (1.4, 12.8)	48.2 (10.2, 86.2)
45–				
Non-smokers	52.4 (4)	11.4 (1)	4.6 (0.6, 33.6)	40.9 (−12.7, 94.6)
Smokers	206.7 (17)	27.9 (2)	7.4 (2.1, 25.7)	178.8 (67.4, 290.2)

Table VI *Deaths from circulatory disease (ICD 390–458) among ever-users of oral contraceptives and controls, by parity**

| | Mortality rate (no. of deaths) | | | |
| | | | | |
Parity	Ever-users	Controls	Relative risk (95% confidence limits)	Excess risk (95% confidence limits)
0	0.0 (0)	10.0 (3)	0.0	−10.0 (−108.0, 88.0)
1–2	22.7 (20)	5.5 (5)	4.2 (1.5, 11.6)	17.2 (5.1, 29.5)
3+	42.8 (35)	5.7 (2)	7.5 (2.8, 19.8)	37.1 (19.4, 55.6)
Total	29.9 (55)	7.2 (10)	4.2 (2.3, 7.7)	22.7 (13.2, 32.2)

*Standardised for age, smoking, and social class.

Table VII *Circulatory disease (ICD 390–458) mortality rates per 100 000 woman-years by total duration of oral contraceptive use*

	Controls	Total duration of oral contraceptive use (mo)				
		1–24	25–48	49–72	73–96	97+
Rate (no. of deaths)	7.0 (10)	32.5 (5)	23.5 (4)	28.4 (5)	32.3 (5)	20.4 (4)
Relative risk	1.0	4.6	3.4	4.1	4.6	2.9

*Standardised for age, parity, and smoking status.

to the large difference between the rates of controls and of current-users in the 1–24-month group.

Because we were uncertain whether the aggregation of intermittent periods of oral contraceptive use might have concealed a trend related to continuous duration of use, we performed the same analysis in women who had taken the pill without a break. The results were similar. The rates for continuous takers were higher than for controls, but there was no evidence of an effect of duration of use.

DISCUSSION

Ever-users of oral contraceptives in this study had a mortality rate from circulatory diseases 4.2 times greater than that of controls. A wide variety of diseases of the circulatory system occurred more commonly in ever-users than in controls. However, only two specific causes (ischaemic heart disease and subarachnoid haemorrhage) were significantly more frequent in ever-users than in controls, and they account for most of the deaths attributed to circulatory diseases.

The relative risk of circulatory mortality from ever-use of the pill increased with increasing age and was greater in smokers than in non-smokers. Older women who smoked and had used the pill were at much higher risk than their non-smoking controls.

The underlying death rate for the general population from circulatory disease increases with age. Therefore, the higher relative risk in older women, if they smoke and use oral contraceptives, results in a substantial increase in their absolute excess risk. For example, ever-users aged 45 and older who smoke have an excess risk of 195.3 per 100 000 woman-years (or one excess death per 512 women per year) compared with their non-smoking controls.

It should be noted that our data on smoking habits were obtained only on entry to the study in 1968–69. Thus there are probably now some non-smokers in the group of women which we categorise as smokers. This will have the effect of underestimating the effect of smoking on death from vascular causes. It is less likely that our group of non-smokers now includes any important proportion of smokers, so that our estimate of risk associated with oral contraceptive use in non-smokers is unlikely to have been materially affected by a change in smoking habits since recruitment to the study.

The demonstration of an increasing risk of mortality from circulatory disease in pill users as parity increases is new, and therefore requires confirmation in other studies. This relationship was evident after standardisation of age, smoking, and social class, and is therefore independent of these variables.

There is now no evidence that duration of oral contraceptive use is associated with the death rate from circulatory disease independently of the increasing risk with age. This observation is contrary to our earlier findings, but agrees with those of others.[8, 10]

An important question is whether the association of oral contraceptive use with circulatory mortality continues after cessation of pill use. That the mortality rate is raised in former-users (table III) does not necessarily imply a residual effect. For example, a woman may have a myocardial infarction while still on the pill, stop taking it, and die later with the death being ascribed to myocardial infarction. In our analysis she would be classified as a former-user at the time of death, although the illness resulting in her death actually began while she was a current-user. This and similar sequences of events would result in an over-estimation of the death rate in former-users and an underestimation of the death rate in current-users. Such cases do not pose serious problems in our analyses since they can be readily identified, and the deaths reassigned to the appropriate user-group.

Unfortunately, in other circumstances it is difficult to determine whether the circulatory death in a former-user was the result of a condition that developed while she was on the pill. For example, if hypertension or headaches developed in a woman taking the pill, she stopped taking it, and she later died of a cere-brovascular disease, it is debatable whether the underlying morbid event occurred when she was a current-user or a former-user. Because of these difficulties, we have more confidence in the mortality rates for ever-users than in the rates for current or former users, and have compared ever-users with controls for most of our analyses.

Nevertheless, the risk of death in former-users is an important issue. While we cannot precisely quantify this risk, at least some statements can be made. For non-rheumatic heart disease and hypertension four of the ten deaths among former-users were in women who acquired their ultimately fatal condition while still taking the pill. If these are regarded as occurring in current-users then the remaining former-users had little, if any, increased risk of death from these conditions compared with women who had never used the pill.

In contrast, the risk of cerebrovascular disease in former-users is less certain. Only 1 of 19 former-users who died of these causes had a stroke while taking the pill. If that one death is regarded as occurring in a current-user, then former-users still have an increased risk of cerebrovascular disease. Of the 18 remaining former-users, 10 acquired conditions while on the pill, which might have been related to stroke—headaches, hypertension, and suspected deep vein thrombosis (DVT). However, these conditions were stated as the reasons for discontinuing pill use in only 40% (4/10) of these cases, and only in the case of the suspected DVT was there an indication that the condition was regarded as serious. Furthermore, there was no clustering of deaths soon after stopping the pill. In the 10 former-users who had possibly related symptoms, only 1 (10%) died in the first year after stopping oral contraceptives; of the 8 former-users without such symptoms 2 (25%) died within a year of stopping. This suggests that the previous morbidity was not closely linked in time with the subsequent deaths, and implies that the raised death rate from cerebrovascular disease in former-users cannot be entirely attributed to events initiated during previous oral contraceptive use.

An important advantage of a prospective cohort study is that a direct estimate of the excess risk of death associated with oral contraception can be obtained. Extrapolations from such estimates are hazardous since women in this study tend to be healthier and have lower death rates that the general population.[1] Further, pill users in our cohort tend to be older than pill users in the general population, and most of the data are derived from users of pills containing $50 \mu g$ of oestrogen. Nevertheless, the estimates help to discriminate women with high and low risks associated with pill use, as well as the relative contribution of different causes of death to the increased mortality in pill users. As mentioned earlier, the excess risk of death from vascular diseases in ever-users of 22.7 per 100 000 woman-years accounts for virtually all of the total excess risk for ever-users of 23.3 per 100 000 woman-years. The excess of vascular causes of death is largely due to cerebrovascular disease and non-rheumatic heart disease and hypertension (9.7 per 100 000 woman-years for each category). Despite the well-known relationship between pill use and thrombophlebitis and pulmonary embolism,[8] these causes account for only about 10% of the excess risk among ever-users of the pill (2.5 per 100 000 woman-years).

The possibility that bias accounts for the reported increased risk of circulatory disease in ever-users must be considered. As in any observational study, there could theoretically be a selection of unhealthy women into the ever-user group. In fact, on the contrary, doctors have tended to advise against the use of the pill in women with significant illness. In this study controls had more diabetes, cerebrovascular disease, and congenital heart disease than ever-users at recruitment.[2] More ever-users smoked, but this was adjusted for in our analyses. While selection factors may have produced other small differences between ever-users and controls, they are unlikely to explain the relative risk of over 4 for ever-users that we found for circulatory disease.[9]

Another possible bias is that, because of their knowledge of the hazards of oral contraception, doctors are more likely to attribute deaths in ever-users to circulatory causes. If false attribution of deaths in ever-users to circulatory causes accounted for the association of pill use with circulatory mortality, there should be a deficiency of deaths from non-circulatory causes in the ever-users. However, the death rates from non-circulatory causes were almost identical in ever-users and controls.

Another possibility is that differential loss to follow-up between the ever-users and controls accounts for the association of pill use with circulatory death. The average annual losses to follow-up are similar for ever-users and controls (6.4% per year and 6.6% per year, respectively). The losses are primarily due to women moving home and, thus, changing their general practitioners, and to the 2.4% average annual loss of participating doctors, mostly because of retirement. These losses from the study population are unlikely to be related to the patients' morbidity or contraceptive use.

In general, our results support the data from case-control studies of the association of oral contraceptive use with ischaemic heart disease.[8, 10] For sub-arachnoid haemorrhage, the results from other studies are conflicting. Petitti and Wingerd[11] found a relative risk for subarachnoid haemorrhage of 6.5 for takers and 5.3 for ex-takers of oral contraceptives, though these estimates were based on a total of only 11 cases. Inman[12] conducted a large case-control study of fatal subarachnoid haemorrhage in women 15–44 years old. He found a relative risk of about 1.5 associated with pill use, which was not a statistically significant increase. In the American study by the Collaborative Group for the Study of Stroke in Young Women,[13] a relative risk of 2.0 was found for haemorrhagic stroke. The proportion of these due to sub-arachnoid haemorrhage was not stated, but most were probably from this cause. In the Oxford/Family Planning Association study there was no excess of subarachnoid haemorrhage in oral contraceptive users, though the excess of deaths due to total vascular diseases was compatible with our estimates.[14, 15]

We cannot explain the difference between the results of these studies and our own. The reported cases are unlikely to have been misdiagnosed. It is possible that the different results reflect predominant use of oral contraceptives with differing oestrogen and progestogen activity[16-18] or the differing prevalence of other risk factors, such as smoking habit. Perhaps the most likely explanation is that the different estimates of risk arise from the sampling errors in the respective study populations, since the 95% confidence limits of all the estimates overlap substantially.

CONCLUSIONS

The findings in this report are not independent of our earlier mortality report[1] since 101 of the 249 deaths in this analysis were included in the earlier paper. Thus, a comparison of the findings between the two reports is inappropriate and the present report should be regarded as superseding the previous one. On the other hand, the clinical implications[19] of the differences between the two reports are important and need to be considered in relation to the observations from other studies.

The main difference between the two reports is that we no longer find an increasing risk of death due to vascular disease with increasing duration of pill use. This difference may be due to several factors. Firstly, the greater number of deaths has permitted us to divide more finely periods of duration of use. In the previous analysis only controls and women with continuous pill usage of less than 5 years, or 5 years or longer were considered. Secondly, we have now included only the mortality rates among pill users in testing for a duration-of-use effect and have excluded the low circulatory mortality rates in controls. Thirdly, in the present duration-of-use analysis we standardised for smoking and parity as well as age. In conformity now with the results of other studies,[8, 10] our data suggest that the widespread belief that women should take into account the length of time they have used oral contraception, in deciding whether to continue pill use, needs reconsideration.

For women under 35 years of age our estimated excess risk associated with pill usage is small and could be due to chance: 1 per 77 000 women per year in non-smokers and 1 per 10 000 women per year in smokers. For women 35-44 years of age the excess risk is greater: 1 per 6700 women per year in non-smokers (95% confidence limits: −1 in 30 000 to +1 in 3000) and 1 per 2000 women per year in smokers (1 in 9800 to 1 in 1200). These risks are compatible with the estimates made in previous case-control studies.[8, 10] These estimates permit a more flexible approach to women who are between 35 and 45 years of age. It is now apparent that the major risk occurs in smokers. Some non-smokers in this age group who have no other risk factors for vascular disease might find the benefits of oral contraception outweigh the estimated risk, particularly if duration of use of the pill is less important than we had previously believed. There is likely to be a further lowering of risk associated with the use of oral contraceptive preparations characterised by the lowest oestrogen and progestagen activity consistent with contraceptive effectiveness.[17, 18] For women 45 years of age and above, we believe that use of oral contraceptives can be justified only in exceptional circumstances.

Our observation of the association of risk with parity needs confirmation. The potentially worrying observation of an increased risk associated with previous use of the pill, which we noted also in our previous analysis, cannot be properly interpreted from these mortality data. It is hoped that an analysis of first episodes of total vascular diseases (fatal and non-fatal), which we are now undertaking, will clarify this and other important issues.

ACKNOWLEDGEMENTS

We thank the 1400 general practitioners who are contributing all the data for this survey. The study is supported by a major grant from the Medical Research Council. The costs of the pilot trials and current supplementary expenditure have been met by the Scientific Foundation Board of the Royal College of General Practitioners. The Board gratefully acknowledges the receipt of funds for research into oral contraception from Organon Laboratories Ltd, Ortho Pharmaceutical Corporation, Schering Chemicals Ltd, G. D. Searle and Co Ltd, Syntex Pharmaceuticals Ltd, and John Wyeth and Brother Ltd.

REFERENCES

1. Royal College of General Practitioners' Oral Contraception Study. Mortality among oral contraceptive users. *Lancet* 1977; **ii**: 727–31.
2. Royal College of General Practitioners. Oral contraceptives and health. London: Pitman Medical, 1974.
3. Wingrave, S. J., Beral, V., Adelstein, A. M., Kay, C. R. Comparison of cause of death coding on death certificates with coding in the Royal College of General Practitioners' Oral Contraception Study. *J Epidemiol Comm Health* (in press).
4. International Classification of Disease. Manual of the International Statistical Classification of Diseases, Injuries, and Causes of Death. Geneva: World Health Organization, 1967.
5. Macmahon, B., Pugh, T. F. Epidemiology: Principles and methods. Boston: Little, Brown, 1970.
6. Miettinen, O. S. Estimability and estimation in case-referrent studies. *Am J Epidemiol* 1976; **103**: 226–35.
7. Mantel, N. Chi-square tests with one degree of freedom, extensions of the Mantel-Haenszel procedure. *J Am Stat Assoc* 1963; **58**: 690–700.
8. Vessey, M. P., Mann, J. I. Female sex hormones and thrombosis. *Br Med Bull* 1978; **34**: 157–62.
9. Lilienfield, A. M. Foundations of epidemiology. New York, Oxford: Oxford University Press, 1976.
10. Shapiro, S., Sloane, D., Rosenberg, L., *et al.* Oral contraceptive use in relation to myocardial infarction. *Lancet* 1979; **i**: 743–47.
11. Petitti, D. B., Wingerd, J. Use of oral contraceptives, cigarette smoking, and risk of subarachnoid haemorrhage. *Lancet* 1978; **ii**: 234–36.
12. Inman, W. H. W. Oral contraceptives and fatal subarachnoid haemorrhage. *Br Med J* 1979; **ii**: 1468–70.
13. Collaborative Group for the Study of Stroke in Young Women. Oral contraception and increased risk of cerebral ischemia or thrombosis. *N Engl J Med* 1973; **288**: 871–78.

14. Vessey, M. P., McPherson, K., Johnson, B. Mortality among women participating in the Oxford/Family Planning Association Contraceptive Study. *Lancet* 1977; **ii**: 731-33.

15. Vessey, M. P., McPherson, K., Yeates, D. Mortality in oral contraceptive users. *Lancet* 1981; **i**: 549.

16. Inman, W. H. W., Vessey, M. P., Westerholm, B., Engelund, A. Thromboembolic disease and the steroidal content of oral contraceptives. A report to the Committee on Safety of Drugs. *Br Med J* 1970; *ii*: 203-09.

17. Kay, C. R. The happiness pill? *J Roy Coll Gen Pract* 1980; **30**: 8-19.

18. Meade, T. W., Greenberg, G., Thompson, S. G. Progestogens and cardiovascular reactions associated with oral contraceptives and a comparison of the safety of 50- and 30-μg oestrogen preparations. *Br Med J* 1980; **280**: 1157-61.

19. Kuenssberg, E. V., Dewhurst, J. Mortality in women on oral contraceptive. *Lancet* 1977; **ii**: 757.

9 Multi-practice research: a randomized controlled trial

Bonnie Sibbald

INTRODUCTION

Hypertension is one of the most common and challenging of conditions managed principally by general practitioners. Its management has been constantly subject to review and revision but remains strongly influenced by the findings of the MRC Mild Hypertension Study reported in 1985. This study was among the largest of clinical trials ever to be conducted within British general practice and exemplifies many of the strengths and weaknesses of controlled trial methodology. It is therefore valuable to examine the MRC trial critically and ask whether the study's design, execution, and findings justify its impact on subsequent clinical practice. Readers who wish to learn more about controlled trial methodology may find it helpful to consult Pocock (1983).

THE STUDY

The MRC Mild Hypertension Study was initiated in 1977 in order to determine whether drug treatment of mild hypertension might reduce the rates of stroke, death due to hypertension, coronary events, and/or overall mortality in men and women aged 35–64 years. Its subsidiary objectives were to compare two active drug treatments (bendrofluazide and propanolol) in terms of cardiovascular outcomes and the nature and incidence of suspected adverse reactions. The study was designed as a randomized, single blind, placebo controlled trial in which an estimated 18 000 patients were to be followed for a period of 5 years each. Recruitment and follow-up were carried out almost exclusively within general practice. The study cost an estimated £2 million spread over a period of eight to ten years.

THE CONTEXT

The first question to ask is whether such a massive and costly trial was necessary.

By the 1970s, large observational studies had established that the risks of stroke and coronary heart disease decline with each successively lower level of blood pressure (Kannel *et al.* 1970). It was therefore logical to postulate that the benefits of reducing blood pressure might be considerable, even in people with mild hypertension. However, observational studies cannot prove that the link between a risk factor and a disease is causal. In order to see whether reducing pressure would in fact reduce the incidence of cardiovascular disease, placebo controlled trials of antihypertensive treatment were needed.

At the outset of the MRC Mild Hypertension Study only one large placebo controlled trial of antihypertensive treatment had been reported. This was the Veterans Administration Cooperative Study (1967) in which 523 middle-aged men were randomly assigned to treatment with either hydrochlorothiazide-reserpine, hydralazine hydrochoride, or placebo and followed for an average period of 5 years. The findings showed that, in men with initial diastolic pressures of 115 to 129 mm Hg, treatment was associated with a significant reduction in stroke and stroke recurrence. However, in men with initial diastolic pressures of 90 to 114 mm Hg, the findings were inconclusive. Treatment was associated with a reduction in stroke but the benefits were confined principally to those with diastolic blood pressures above 104 mm Hg and those aged over 50 years (Veterans Administration Co-operative Study Group 1970). With the benefit of hindsight it was clear that the study had been too small adequately to evaluate treatment in mild hypertension.

Thus by the mid 1970s the value of treating severe hypertension was established, while the value of treating mild to moderate hypertension remained unproven. No studies in women had yet been reported, and the impact of antihypertensive treatment on coronary heart disease had not been elucidated. Many clinicians, inspired by the theoretical rather than the proven benefits of therapy, undertook to treat patients with diastolic pressures as low as 95 mm Hg (World Health Organisation 1962). The need to establish the efficacy of drug treatment in mild hypertension was urgent and many countries responded by setting up studies to address this issue. In Australia a large placebo controlled trial of 'stepped care' was initiated (Australian National Blood Pressure Study Management Committee 1980). In the USA the Hypertension Detection and Follow-up Program (1979) was set up to compare 'specialist' management with 'usual' management. Britain was among the last to respond by initiating the MRC Mild Hypertension Trial in 1977.

It is reasonable to ask why British clinicians initiated their own study rather than awaiting the outcome of others. The answer lies in the considerable differences among countries both in the nature and organization of primary care and in doctors' preferred management strategies. The MRC trial was conducted within British general practice using the drug treatments most commonly prescribed by British GPs. As such it was designed to have maximum relevance to the clinicians who must act upon its findings.

STUDY DESIGN

A pilot study was conducted to establish the feasibility of the main study and to resolve any difficulties with its administration (MRC 1977). This is a valuable preliminary step to any investigation, but is essential in the planning of very large and/or expensive studies.

The random assignment of patients to treatment groups is an essential aspect of controlled trials in that it ensures that differences in outcome are not attributable to any systematic differences among groups at the outset. In the MRC study patients were randomly allocated to one of four treatment groups: bendrofluazide (10 mg daily); placebo bendrofluazide; propanolol (up to 240 mg daily); and placebo propanolol. Randomization was in blocks of eight by sex, 10 year age band, and study centre. This means that for every eight patients recruited in each age–sex band in each study centre there would two patients in each of the four study groups. The method of randomizing within 'blocks' ensures that treatment groups will be comparable with respect to the parameters defining the block.

The selection of drugs and dosages was based on those most commonly used by GPs at the outset of study. However at its conclusion, new potentially more appropriate drugs had been developed and the recommended dosage of diuretic had been reduced substantially. The relevance of the findings was therefore diminished – a common problem in any long term study where clinical practice changes constantly.

The study design permitted supplementary drug treatment to be given to any patient on active treatment whose diastolic pressure was not reduced below 90 mm Hg by six months. In all treatment groups, any appropriate medical intervention was allowed if pressure was sustained at or above 110 mm Hg diastolic or 200 mm Hg systolic or both. In the event, two fifths of patients lapsed from the treatment to which they were originally assigned. Such high levels of treatment alteration and withdrawal complicate interpretation of the study findings, but were necessary if the trial was to be ethical. One of the principal limitations of experimental research in human beings is that the study design must often be compromised by the need to safeguard the health and well being of subjects.

Because dosage adjustment was a necessary component of the study design, the investigators thought it necessary to inform doctors about which treatment a patient was taking. The study was therefore conducted 'single blind' and open to observer bias by doctors. The bias was offset by selecting procedures for the recording of blood pressure and outcomes which were least sensitive to interpretation by doctors (see below). It remains possible that patients on placebo were less likely than those on active treatment to be withdrawn because of an apparent side effect. The study is therefore likely to have overestimated the incidence and severity of adverse reactions in patients on active treatment.

STUDY POPULATION

The aim was to recruit 18 000 men and women aged 35–64 years of age with a sustained phase V diastolic blood pressure of 90–109 mm Hg together with a systolic pressure below 200 mm Hg. This large sample was needed to give a 95 per cent chance of detecting a 40 per cent reduction in stroke which would be significant at the 1 per cent level. The calculations were based on the rates of stroke reported in previous epidemiological studies and the Registrar General's mortality statistics. The statistical power and level of significance were higher than is usual and presumably reflected the conviction that such a trial was unlikely ever to be repeated and should therefore be decisive.

It is questionable whether diastolic pressure should have been used in preference to systolic pressure in the recruitment and monitoring of patients. Published evidence at that time and subsequently suggests that systolic pressure is a better predictor of stroke and coronary heart disease than diastolic pressure (Rutan *et al.* 1989). The focus on diastolic pressure appears to have originated with the Veterans Administration Study which established a precedent for subsequent trials. In following this precedent, the MRC trial ensured that the findings could readily be compared with those of other studies.

The recruitment and follow-up of patients was conducted largely within general practice. This demonstrates both the feasibility of using practice lists to identify general population samples and the ability of GPs to work effectively together in addressing problems which are not amenable to study within individual practices. Most importantly, by conducting the study within the general practice setting, the investigators maximized the study's relevance to the clinicians principally responsible for hypertension management.

The network of practices recruited to this investigation formed the basis of the modern GPRF (General Practice Research Framework) which is maintained by the MRC based at St Bartholemews Hospital in London. The framework offers excellent opportunities for health services research but has been little used by GPs to address questions specific to their work. It is hoped that in future the GPRF will be utilized more often by GPs on behalf of GPs.

Participating practices within the MRC mild hypertension study had to provide sufficient space and staff for the study and there was a bias towards large group practices in small towns whose patients were predominantly white and from the upper socio-economic groups. The study population was therefore not representative of the general population, affecting generalizability. This bias will not have distorted the comparisons made between treatment groups. However, as compliance with drug treatment and cardiovascular morbidity vary with people's socio-economic, ethnic, and racial characteristics, the outcomes of this trial cannot readily be extrapolated to all sectors of the British population.

DATA COLLECTION AND QUALITY CONTROL

The investigators took all reasonable precautions to ensure that the study data were reliable and reproducible. All important terms were defined precisely thereby minimizing ambiguity in interpretation and ensuring consistency. Most data were based on objective measurements, not the clinical judgements or opinions of doctors. Random zero sphygmomanometers were used to minimize bias in blood pressure recording. The cause of death was determined by an independent expert who was blind to the patient's treatment group and made decisions based on all available clinical information. Save for the possible bias introduced by the single blind design (see above), all study groups appear to have been treated comparably. This is important in that any systematic differences in the monitoring of groups would undermine their comparability and could lead to false conclusions regarding the value of active treatment.

In well-managed trials, random checks are carried out to ensure that the active and placebo treatments have not accidentally been confused during production and packaging. No mention is made by the investigators on this point.

In all clinical trials it is important to minimize loss to follow-up and so maintain both the comparability and representativeness of study groups. In this investigation 20 per cent of patients of both sexes in each study group were lost to follow up. This rate of attrition is not unusual in clinical trials and may reduce the validity of comparisons between study groups and inferences extrapolated from the study population to the patient population as a whole.

DATA PRESENTATION AND ANALYSIS

The main and subsidiary findings are presented clearly, objectively, and in sufficient detail to enable readers to judge for themselves. It is necessary, however, to read this main paper in conjunction with previous (MRC 1977,

1981) and subsequent (MRC 1988) reports in order to obtain a full picture of the study design and outcomes. This is not unreasonable in that inclusion of these additional data within the main paper would have produced a report of overwhelming length and complexity.

The significant feature of the data analysis is that it was by 'intention to treat'. This means that all patients recruited to a particular treatment group are included with that group in the analysis, regardless of whether they remained on their original treatment. The strength of an intention to treat analysis is that it maintains the randomization essential to the controlled trial design. We can be confident that there are no systematic differences among treatment groups, save for the treatment itself, which could explain the outcomes. The important limitation of such analyses is that they usually provide a conservative estimate of the likely benefits of treatment. In other words the analysis is likely to underestimate the true treatment effect.

In general, clinicians also want to know the outcome of 'on treatment' analyses – what happens to the patients who are fully compliant and their treatment unaltered. Such patients always form a selected subgroup which is unrepresentative both of the initial treatment group and of the patient population as a whole. On treatment analyses are therefore unsafe in that we cannot be confident that apparent differences in outcome are attributable only to the differences in treatment. The tendency is to overestimate the likely benefits of treatment because failures are excluded. Despite these drawbacks, clinicians would have been interested to see the results of the on treatment analysis in order to gain some appreciation of the outcome in fully treated patients. The investigators state simply that 'there were only minor differences between the results of the intention to treat and on treatment analyses'. This finding is surprising and deserves further explanation given that approximately two fifths of patients lapsed from their original treatment during the course of the study.

Withdrawal from treatment was occasioned mainly by adverse drug reactions in patients receiving active medication and by inadequate blood pressure control in those on placebo. The principal adverse reactions reported with bendrofluazide were impaired glucose tolerance in both sexes (7.7 males and 5.9 females per 1000 patients per year), and in men, gout (12.8 per 1000 patients per year) and impotence (12.6 per 1000 patients per year). Those associated with propanolol were dyspnoea (5.3 males and 8.3 females per 1000 patients per year) and Raynaud's phenomenon (5.1 males and 4.5 females per 1000 patients per year) in both sexes, and again impotence (6.3 per 1000 patients per year) in men. It is a tribute to the size and duration of this study that it was able to establish the incidence of long term side effects for these two common treatments.

The investigators include the results of *post hoc* analyses aimed at identifying which subgroups of patients benefited most from treatment. They rightly emphasize the need to be 'very cautious' in judging the findings. The study was not designed for this purpose and may lack the necessary statistical power. This is because, in searching the data for likely factors to explain outcomes, it is necessary to make many comparisons which increases the likelihood that some will turn out statistically significant by chance alone. For these reasons the findings of *post hoc* analyses should always be viewed, not as proven outcomes, but as hypotheses which require further testing.

THE CONCLUSIONS

The principal conclusion was that active treatment was significantly able to reduce the risk of stroke but not of coronary events or overall mortality. There was no overall advantage of one active treatment over the other, and the benefits were offset by the high prevalence of side effects, not all of which were mild or trivial. These conclusions were relevant to the stated objectives, consistent with the data, and not extrapolated beyond the scope of the research findings.

The GP reaction was predictably one of disappointment. The prevalence of side effects was much higher than previously supposed, particularly in respect of bendrofluazide which previously was not known to cause impotence. The benefits of treatment were far smaller than expected. Approximately 850 patients would need to be treated for one year to prevent one stroke. Lives would not be saved and there was no appreciable impact on the incidence of coronary events.

This disappointment was offset in part by the results of *post hoc* analyses which offered more specific guidance on which types of patient might benefit most from which types of treatment (MRC 1988). The results suggested that clinically meaningful reductions in stroke could be achieved by bendrofluazide in smoking men aged 55–64 years with elevated systolic pressures, and by propanolol in non-smoking men of comparable age and systolic pressure. The recommendations for women were similar although the benefits in terms of stroke reduction were less marked.

These findings greatly influenced guidance on the treatment of mild hypertension issued by the British Hypertension Society Working Party (1989) and reiterated in its recent guidance (Sever *et al.* 1993). It was accepted that not all patients with mild to moderate hypertension warranted treatment. Antihypertensive drugs should instead be reserved for patients at high risk of stroke or coronary heart disease by virtue of other factors such as their age, sex, and/or existing end organ damage. While this advice accurately reflected the research evidence, it is important to recall that this evidence was insecure having been derived from the *post hoc* analyses of a trial designed for a different purpose. The findings should have been subject to verification in a second placebo controlled trial. In practical terms this was not possible, in part because of the great financial cost, but principally because it would be viewed as unethical to withhold treatment from the 'high risk' patients in a placebo group. This again demonstrates how ethical considerations place necessary boundaries on scientific enquiry.

Apart from its contribution to clinical management, the MRC trial was able to add important information to a growing body of research on the epidemiology of cardiovascular disease. The findings confirmed and refined those of other studies in showing that the entry characteristics which predicted subsequent stroke were elevated systolic or diastolic blood pressure, age, male gender, and cigarette smoking. Pre-existing ischaemic heart disease, elevated serum cholesterol, and high body mass index were associated with an increased risk of coronary events, but not stroke.

Particular attention was drawn to the finding that smoking at entry had a bigger effect on morbidity and mortality than did antihypertensive treatment. It was tempting to suggest that stopping patients smoking would achieve more substantial benefits than would blood pressure reduction. However, it would be wrong to draw such a conclusion from this study alone. Although smoking was clearly an important risk factor, it does not necessarily follow that stopping smoking would substantially reduce the incidence of stroke or coronary heart disease. Other research was needed to establish the value of smoking cessation (Shaper *et al.* 1991).

THE VERDICT

The MRC Mild Hypertension Study was skilfully designed, properly executed, and well presented. Its findings showed that the adverse consequences of drug treatment in mild hypertension may, for many patients, outweigh the benefits of treatment. Much of course remained to be learned. Subsequent studies have examined the management of hypertension in the elderly and work continues into the role of newer antihypertensive agents including angiotensin enzyme inhibitors, calcium channel blockers, and alpha blockers (Sever *et al.* 1993).

One problem of central importance to all such work is how to balance the relative costs and benefits of therapy in the management of an asymptomatic condition such as hypertension. Strategies which benefit the health of populations do not always maximize the quality of life for individuals. If patients are to make informed decisions, they need adequately to be appraised of both the relative and absolute risks of ill health with and without treatment. Yet little research has been conducted into how GPs convey such complex aspects of risk to patients and what factors influence patients' decisions. These are issues which deserve to be the focus of more in depth research.

REFERENCES

Australian National Blood Pressure Study Management Committee (1980). The Australian therapeutic trial in mild hypertension. *Lancet*, **i**, 1261-7.

British Hypertension Society working party (1989). Treating mild hypertension: agreement from the large trials. *British Medical Journal*, **298**, 694-8.

Hypertension Detection and Follow-up Program Cooperative Group (1979). Five-year findings of the hypertension detection and follow-up program. (1) Reduction in mortality of persons with high blood pressure, including mild hypertension. (2) mortality by race, sex, and age. *JAMA*, **242**, 2562-77.

Kannel, W. B., Wolf, P. A., Verter, J., and McNamara, P. M. (1970). Epidemiological assessment of the role of blood pressure in stroke. *JAMA*, **214**, 301-10.

Medical Research Council Working Party on Mild to Moderate Hypertension (1977). Randomised controlled trial of treatment for mild hypertension: design and pilot study. *British Medical Journal*, **1**, 1437-40.

Medical Research Council Working Party on Mild to Moderate Hypertension (1981). Adverse reactions to bendrofluazide and propanolol for the treatment of mild hypertension. *Lancet*, **ii**, 539-43.

Medical Research Council Working Party (1985). MRC trial of treatment of mild hypertension: principal results. *British Medical Journal*, **291**, 97-107.

Medical Research Council Working Party (1988). Stroke and coronary heart disease in mild hypertension: risk factors and the value of treatment. *British Medical Journal*, **296**, 1565-70.

Pocock, S. J. (1983). *Clinical trials: a practical approach*. John Wiley & Sons, Chichester.

Rutan, G. H., McDonald, R. H., and Kuller, L. H. (1989). A historical perspective of elevated systolic vs. diastolic blood pressure from an epidemiological and clinical trial viewpoint. *Journal of Clinical Epidemiology*, **42**, 663-73.

Sever, P., Beevers, G., Bulpitt, C., Lever, A., Ramsay, L., Reid, J., and Swales, J. (1993). Management guidelines in essential hypertension: report of the second working party of the British Hypertension Society. *British Medical Journal*, **306**, 983-7.

Shaper, A. G., Phillips, A. N., Pocock, S. J., Walker, M., and Macfarlane, P. W. (1991). Risk factors for stroke in middle aged British men. *British Medical Journal*, **302**, 1111-15.

Veterans Administration Co-operative Study Group (1967). Effects of treatment on morbidity in hypertension. Results in patients with diastolic blood pressure averaging 115 through 129 mm Hg. *JAMA*, **202**, 1028-34.

Veterans Administration Co-operative Study Group (1970). Effects of treatment on morbidity in hypertension. II Results in patients with diastolic blood pressure averaging 90 through 114 mm Hg. *JAMA*, **213**, 1143-52.

World Health Organisation (1962). Arterial hypertension and ischaemic heart disease: preventive aspects. Report of an expert committee. Technical report series no. 231.

MRC trial of treatment of mild hypertension: principal results

Medical Research Council Working Party

BMJ, **291**, 97–104 (1985)

Abstract

The main aim of the trial was to determine whether drug treatment of mild hypertension (phase V diastolic pressure 90–109 mm Hg) reduced the rates of stroke, of death due to hypertension, and of coronary events in men and women aged 35–64 years. Subsidiary aims were: to compare the course of blood pressure in two groups, one taking bendrofluazide and one taking propranolol, and to compare the incidence of suspected adverse reactions to these two drugs. The study was single blind and based almost entirely in general practices; 17 354 patients were recruited, and 85 572 patient years of observation have accrued. Patients were randomly allocated at entry to take bendrofluazide or propranolol or placebo tablets.

The primary results were as follows. The stroke rate was reduced on active treatment: 60 strokes occurred in the treated group and 109 in the placebo group, giving rates of 1.4 and 2.6 per 1000 patient years of observation respectively (p < 0.01 on sequential analysis). Treatment made no difference, however, to the overall rates of coronary events: 222 events occurred on active treatment and 234 in the placebo group (5.2 and 5.5 per 1000 patient years respectively). The incidence of all cardiovascular events was reduced on active treatment: 286 events occurred in the treated group and 352 in the placebo group, giving rates of 6.7 and 8.2 per 1000 patient years respectively (p < 0.05 on sequential analysis). For mortality from all causes treatment made no difference to the rates. There were 248 deaths in the treated group and 253 in the placebo group (rates 5.8 and 5.9 per 1000 patient years respectively).

Several post hoc analyses of subgroup results were also performed but they require very cautious interpretation. The all cause mortality was reduced in men on active treatment (157 deaths versus 181 in the placebo group; 7.1 and 8.2 per 1000 patient years respectively) but increased in women on active treatment (91 deaths versus 72; 4.4 and 3.5 per 1000 patient years respectively). The difference between the sexes in their response to treatment was significant (p = 0.05). Comparison of the two active drugs showed that the reduction in stroke rate on bendrofluazide was greater than that on propranolol (p = 0.002). The stroke rate was reduced in both smokers and non-smokers taking bendrofluazide but only in non-smokers taking propranolol. This difference between the responses to the two drugs was significant (p = 0.03). The coronary event rate was not reduced by bendrofluazide, whatever the smoking habit, nor was it reduced in smokers taking propranolol, but it was reduced in non-smokers taking propranolol. The rate of all cardiovascular events was not reduced by bendrofluazide, whatever the smoking habit, or in smokers taking propranolol but was reduced in non-smokers taking propranolol. The difference between the two drugs in this respect was significant (p = 0.01).

INTRODUCTION

The Medical Research Council's trial of drug treatment of mild hypertension began in 1977, after a successful pilot study.[1] By that time controlled trials had shown that treatment was effective in reducing the incidence of events related to hypertension, such as stroke, in severely hypertensive men,[2] in hypertensive men with phase IV diastolic pressures exceeding 115 mm Hg,[3] and in survivors of strokes[4]; and there was some suggestion that men with phase V diastolic pressures of 90–114 mm Hg might also benefit.[5] There was, however, no definite evidence that drug treatment would reduce the rates of stroke or other cardiovascular events in men with phase V diastolic pressures below 110 mm Hg, and there were no controlled trial data at all on the value of treatment for women with mild hypertension.

Aims

This trial was therefore set up, under the guidance of an MRC working party responsible for all major scientific decisions, to establish whether drug treatment of mild hypertension (phase V diastolic pressure 90–109 mm Hg) would be associated with a 40% reduction in the number of deaths due to stroke (*International Classification of Diseases* (eighth revision) 430–438) and hypertension (ICD 400–404) and in the number of non-fatal strokes (power 95%, significance level 1%). It was appreciated that there would also be large enough numbers of fatal (ICD 410–414) and non-fatal coronary events to assess the effects of treatment on this category.

The two subsidiary objectives were: (*a*) to compare the course of blood pressure in two groups of participants, one taking the thiazide diuretic bendrofluazide and one taking the β blocking agent propranolol; and (*b*) to compare the incidence of suspected adverse reactions to these two drugs.

PATIENTS AND METHODS

Study size

Calculations based on epidemiological data[6] and on the Registrar General's mortality statistics[7] suggested that 18 000 men and women aged 35–64 years, each to be followed up for five years (giving a total of 90 000 person years of observation), would be needed to achieve the main aim of the trial with regard to stroke. This size was likely to be at least adequate for assessing a similar effect of treatment on the rate of coronary events, which were expected to occur more frequently than stroke. It was recognised that even this study size would not permit separate analyses for men and women and was unlikely to permit separate analyses for stroke and coronary event rates in people in the two individual active drug groups. The drug groups were kept separate as far as possible, and the two drugs were only exceptionally used as supplements to one another, so that if the results eventually showed that any comparisons of event rates by randomized drug were feasible these would not be invalidated.

Recruitment and screening

Recruiting took place over nine years from March 1973 to February 1982, starting slowly during the pilot phase and proceeding rapidly from 1977 onwards. The pilot study showed that clinics specially established in general practice were at least as satisfactory as similar clinics based in industrial organisations or in large screening projects. Since mild hypertension is usually managed in general practice, general practice clinics were used for most of the main phase of the study. More practices than could be included applied to take part. One of the factors determining a practice's suitability was the availability of space for screening (carried out sometimes within the practice rooms and sometimes in one of the MRC's mobile screening caravans) and for trial clinics. In consequence disproportionate numbers of practices from areas such as small towns were enlisted. By reducing the participation of practices in inner city areas this selection process has probably affected the social class structure of the trial population, biasing it towards the upper socioeconomic groups. The population screened was almost entirely identified from the age-sex registers of 176 group practices distributed throughout England, Scotland, and Wales: 695 000 invitations to attend for screening were sent out, and 515 000 (74%) were accepted.

The blood pressure criteria for entry to the trial were: at screening, diastolic (phase V) pressures of 90–109 mm Hg together with a systolic pressure below 200 mm Hg. Screening pressure was defined as the mean of four readings taken on two separate occasions and confirmed by the mean of two later readings still in this range ('entry pressure'). A total of 46 350 (90% of those screened) had blood pressures in the trial range; 25 750 (5%), however, had some exclusion factor (secondary hypertension; taking antihypertensive treatment; normally accepted indications for antihypertensive treatment (such as congestive cardiac failure) present; myocardial infarction or stroke within the previous three months; presence of angina, intermittent claudication, diabetes, gout, bronchial asthma, serious intercurrent disease, or pregnancy). Of the 20 600 (4%) eligible, 16 410 (almost 80%) agreed to participate, giving signed informed consent. Together with 944 people identified at other screening centres, this gave the total of 17 354 participants; the follow up period was extended to five and a half years.

Age range

People in the trial age range, 35–64 years, were expected to experience fewer strokes and coronary events than would an older population; but, because the impact of such events may be greater in younger people, the importance of obtaining evidence about the value of antihypertensive treatment in this age group was considered sufficient to outweigh this disadvantage. People aged less than 35 were not recruited because their event rate would have been so low.

Treatment regimens

Patients were randomly allocated at entry to one of four treatments: the thiazide diuretic bendrofluazide; placebo tablets that looked like bendrofluazide; the β blocker propranolol; and placebo tablets that looked like propranolol. The two placebo groups were treated as one in all analyses. Randomization was in stratified blocks of eight within each sex, 10 year age group, and clinic. Thiazide diuretics and β adrenoceptor blocking drugs were selected because, firstly, at the time the trial was designed these were the most commonly used pharmacological agents for treating mild to moderate hypertension and, secondly, it was hoped to show whether the incidence of coronary events would be reduced by β blockade. There are important differences in the metabolic, hormonal, and haemodynamic effects of these two types of drug, and it was hoped that useful comparative data would be collected in the trial. The drugs selected from these groups were bendrofluazide and propranolol. There was already considerable experience of their use, which made it less likely that serious toxicity would be discovered. The doses chosen, 10 mg daily of bendrofluazide and up to 240 mg daily of propranolol, were in common use and were known to be roughly equipotent in terms of their hypotensive effect.

The target level of blood pressure for those randomized to active treatment was diastolic pressure (phase V) below 90 mm Hg, to be reached within six months of entry to the study. Supplementary treatment was added if blood pressure did not respond satisfactorily to the primary drug. Methyldopa was originally used as a supplement to bendrofluazide and guanethidine as a supplement to propranolol, but later methyldopa was used whatever the primary drug. Only exceptionally was one of the primary trial drugs used to supplement the other (the five and a half year cumulative percentage was 5%, 2% of the total patient years of observation).

A small group of patients (288) was randomly assigned at entry to a fifth treatment regimen of observation only, taking no tablets but otherwise adhering to the standard protocol. The course of blood pressure in this group was indistinguishable from that in the placebo group,[8] and the two groups have been merged in the analyses.

The changes in dosage in the propranolol group and the availability of supplementary treatment in both actively treated groups sometimes necessitated several adjustments of medication in patients whose blood pressure did not easily reach the target level. When the protocol was written it was judged unreasonable to ask general practitioners to undertake such adjustments in a double blind study, and the trial was therefore single blind only.

Doctors were free to use their own judgment in managing obesity and advising on cigarette smoking, exercise, and salt intake, but they were asked to follow a consistent policy for treated and control patients.

Data collection and quality control

The first four screening measurements and the follow up blood pressure measurements were made by specially trained, and regularly tested, nurses. Confirmatory blood pressure measurements in the later stages of screening and full medical examinations at entry and each year of the trial were performed by the general practitioners. Hawksley random zero sphygmomanometers[9] were used for almost all blood pressure measurements; in only two clinics were London School of Hygiene sphygmomanometers[10] used instead. All forms were checked at the coordinating centre (based in the MRC Epidemiology and Medical Care Unit, Northwick Park Hospital, Harrow), and adherence to the protocol was monitored.

Withdrawal from randomly allocated treatment

Control patients whose blood pressure rose to levels at which placebo treatment was judged unethical were transferred to the corresponding active drug. The original criteria for transfer were a diastolic pressure of 115 mm Hg or a systolic pressure of 210 mm Hg, or both, at two consecutive or three non-consecutive follow up visits. In September 1980 these levels were reduced to 110 mm Hg for diastolic pressure and 200 mm Hg for systolic pressure. If people on active treatment developed pressures at these levels their doctors were free to use whatever drugs they selected irrespective of the protocol.

Other reasons for withdrawal from randomized treatment included the development of complications necessitating active treatment and suspected adverse drug reactions. All patients whose treatment was changed were asked to continue to attend for all the follow up examinations.

Termination of participation in trial

Events terminating a patient's participation were: stroke, whether fatal or non-fatal; coronary events, including sudden death thought to be due to a coronary cause, death known to be due to myocardial infarction, and non-fatal myocardial infarction; other cardiovascular events, including deaths due to hypertension (ICD 400–404) and to rupture or dissection of an aortic aneurysm; and death from any other cause. Clinic staff reported these events to the coordinating centre. The records of all patients who suffered non-fatal terminating events and of any others who lapsed from the trial, whatever the reason, were 'flagged' at the Southport NHS central register to ensure notification of death.

Assessment of terminating events

The evidence on which the diagnosis of each terminating event was based was assessed by an arbitrator ignorant of the treatment regimen. All available documentation was reviewed, including copies of general practitioners' notes, hospital impatient or outpatient notes, electrocardiographic recordings, necropsy findings, and death certificates, and full details were almost always obtained. In virtually all cases classification of fatal events used in the trial analyses was based on this detailed information rather than solely on the wording or coding of the death certificate. The arbitrator used WHO criteria [11, 12] for classification. 'Definite' and 'possible' categories of coronary events were combined, as the distinction between these groups depends not only on the nature of the episode but also on the amount of evidence available.

If a patient had a non-fatal event followed by a fatal event in the same category – for example, a non-fatal and then a fatal stroke – only the fatal event was included in the analyses (38 people were in this group). If a person suffered two events in different categories – for example, a non-fatal stroke and then a coronary event – both were included (six people were in this group).

Data for terminating events were regularly reviewed by the monitoring committee, which prepared reports for the independent ethical committee, whose remit included advising when the trial should stop.

Statistical management

Primary results – All analyses presented here are based on randomized treatment ('intention to treat') categories. Thus data for all participants are presented as if the individual was still in the treatment group to which he was originally randomized, although substantial percentages of patients (see below) were in fact withdrawn from their randomly allocated regimen during follow up. This method of analysis is the preferred approach for randomized clinical trial data, despite the inevitable contamination of the original treatment groups with results for individuals receiving alternative treatments. Although intention to treat analysis may underestimate effects associated with treatment, it is unlikely to lead to false conclusions due to subsequent selection of the group of patients who remain on a particular treatment regimen. This method may fail to detect the consequences of pharmacological effects present only at the time a drug is taken. In fact, there were only minor differences between the results of the intention to treat and on treatment analyses.

Data for terminating events were analysed sequentially. Results were tested every six months. The incidence of the main endpoints of the trial (strokes, coronary events, all cardiovascular events, and all cause mortality) in the two actively treated groups together was compared with that in the two placebo groups together and tested at a stringent nominal p value (using a χ^2 test without continuity correction) allowing the maintenance of an overall type I (false positive) error rate of 0.01 for the beneficial effects of active treatment with 15 analyses of the data.[13] The corresponding overall type I error rate for the adverse effects of treatment was kept at 0.05. Data for comparisons between the two sexes and between the two active regimens were also kept under review.

Secondary results—The results of some post hoc subgroup analyses are presented, although such analyses require very cautious interpretation and can be misleading. Those discussed here are biologically plausible, which helps to substantiate their credibility. The results of these analyses have been presented whether or not they reached conventional or arbitrary levels of statistical significance. Some p values are indeed conventionally significant, but, in ascribing importance to these, the large number of comparisons made must be borne in mind.

Predictive characteristics—Logistic regression analysis was used to estimate the relation between treatment, certain entry characteristics, and the probability of subsequently having a terminating event. The entry characteristics considered were: age, sex, cigarette smoking, ischaemic changes on the electrocardiogram (Minnesota codes 1_{1-2}, 4_{1-3}, 5_{1-2} (one or more)), systolic and diastolic blood pressure, serum cholesterol concentration, and Quetelet's body mass index (body weight/height^2 in kg/m^2). The terminating events examined in these analyses were strokes, coronary events, all cardiovascular events, and all cause mortality. Treatment was considered in two ways: firstly, by comparing data for the active and placebo groups and, secondly, by comparing the results for one primary drug with those for the other. Interaction analyses, using the relation between entry variables and treatment defined in both these ways, were used to assess the importance of possible differences in response to treatment which were found between certain subgroups. Such comparisons were outside the original aims of the trial and may have had only a limited ability to detect even moderate differences. The results in individual subgroups were not subjected to significance testing, as this can often be misleading.[14] The logistic regressions used a controlled stepdown procedure (with some categorisation of continuous variables where necessary because of limitations of computer space). The calculations were performed using generalised linear interactive modelling (GLIM).[15] The p values presented in relation to the results of the logistic regression analyses refer to this 'once off' testing.

All rates shown have been age standardised to the structure of the total trial population.

RESULTS

Numbers, risk factors, and patient years of observation

The numbers of patients recruited, certain entry characteristics, and the patient years of observation are shown in table Im, which confirms that there were no obvious imbalances between the groups in terms of major risk factors at entry. The aim was to accrue 90 000 patient years; in the event 85 572 were achieved by the end of the trial. Closely similar percentages of initially smoking patients in the active and the placebo groups gave up smoking during the trial (24.6% of men on active treatment and 23.4% in the control group of men, with corresponding figures for women of 23.4% and 22.5%). Data for changes in body weight during the trial could not usefully be compared, as each of the two active drugs was itself associated with a change in body weight (a reduction on bendrofluazide and an increase on propranolol) which was significantly different from that in placebo treated subjects. No data for exercise or salt intake were collected.

Course of blood pressure

Average blood pressure fell immediately after entry in all treatment groups (Fig. 1), including those taking placebo tablets and those on observation only. The fall was steepest in the first two weeks; it then continued, more gradually, for about three months. From the first anniversary of entry onwards average pressure changed very little.

Average pressure after entry was lower in those taking bendrofluazide than in those taking propranolol. Within the propranolol group pressure control was less effective in older people; the details have been published.[16] The percentages of participants with diastolic pressure at the target level (below 90 mm Hg) were consistently higher in the bendrofluazide group than in the propranolol group (table IIm). The use of supplementary drugs by those randomized to bendro-

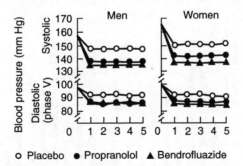

Figure 1. *Mean levels of blood pressure by sex and randomized treatment group.*

Table I *Numbers, patient years of observation completed, and entry characteristics of treatment groups*

	Men				Women			
	Bendrofluazide	Propranolol	Placebos	Pooled SD	Bendrofluazide	Propranolol	Placebos	Pooled SD
Number (%):	2238 (25)	2285 (25)	4525 (50)		2059 (25)	2118 (25)	4129 (50)	
Patient years:	10 945	11 184	22 190		10 274	10 508	20 471	
Mean age (years)	51	51	51	8	53	53	53	7
Mean body weight (kg)	82	81	81	12	70	70	70	13
Mean systolic blood pressure (mm Hg)	158	158	158	16	165	165	165	17
Mean diastolic blood pressure (mm Hg)	98	98	98	6	99	99	98	6
Mean serum cholesterol (mmol/l)	6.3	6.3	6.3	1.0	6.7	6.7	6.7	1.2
Mean serum potassium (mmol/l)	4.1	4.1	4.2	0.4	4.1	4.1	4.1	0.4
Mean serum urate (μmol/l)	382	374	373	68	297	293	293	63
Mean serum sodium (mmol/l)	142	142	142	2	142	142	142	3
Mean serum urea (mmol/l)	5.4	5.4	5.4	1.2	5.2	5.1	5.2	1.2
% Cigarette smokers	32	30	32		27	25	27	
% With left ventricular hypertrophy on ECG*	0.4	0.3	0.4		0.2	0.2	0.4	
% With Q wave abnormalities†	1.0	1.2	1.5		1.7	1.7	1.4	
% With history of stroke	0.8	0.7	0.7		0.7	0.7	0.7	

* 3, 4, 5 on Minnesota code. † 1 on Minnesota code.
Conversion: SI to traditional units—Cholesterol: 1 mmol/ ≈ 38.6 mg/100 ml. Potassium and sodium: 1 mmol/l ≈ 1 mEq/l. Urate: 1 mmol/l ≈ 17 mg/100 ml. Urea: 1 mmol/l ≈ 6 mg/100 ml.

Table II *Percentages of participants with diastolic blood pressure below 90 mm Hg at anniversary visits*

	Men					Women				
Years since joining trial:	1	2	3	4	5	1	2	3	4	5
Bendrofluazide	66	69	70	72	72	71	75	76	79	78
Propranolol	60	65	66	68	71	64	68	69	73	76
Placebos	38	40	42	43	43	42	44	46	49	50

Table III *Cumulative percentages of participants taking supplementary therapy*

	Men					Women				
Years since joining trial:	1	2	3	4	5	1	2	3	4	5
Bendrofluazide	21	26	30	32	34	15	19	21	23	24
Propranolol	12	18	21	23	25	8	12	14	17	19

fluazide (fixed dose) consistently exceeded that by patients randomized to propranolol (titratable dose) (table III). The extent of separation between average pressures in actively treated and control groups is shown in table IV. Annual measurements showed that between one third and one half of all those taking placebo had diastolic pressures below 90 mm Hg; however, different people made up this total at each anniversary. Only 18% (1270) of the 7141 in the placebo group for whom blood pressure measurements at the first three anniversary visits were recorded had diastolic pressures below 90 mm Hg on each of these three occasions; 23% (1657) were in the target range at two of these visits and 27% (1929) at one visit. Only 32% (2285) had no measurements of diastolic pressure below 90 mm Hg at any of these visits.

Altogether 1011 people randomized to placebo treatment and 76 people randomized to active tablets (table V) developed blood pressure above the mild range.

Withdrawals from randomized treatment and lapses from follow up

Numbers and cumulative percentages of people withdrawn from randomized treatment because they developed either suspected adverse reactions to the primary regimen (discussed in detail elsewhere[17]) or levels of blood pressure above the upper limit for the trial are shown in table V and Fig. 2. The protocol for the follow up routine was the same for these people as for those whose treatment was unchanged.

Table IV *Difference (in mm Hg) between mean levels of blood pressure at anniversaries of entry in actively treated people and those on placebo, by sex and drug*

	Men					Women				
Years since entry:	1	2	3	4	5	1	2	3	4	5
Systolic blood pressure										
Bendrofluazide	13	12	13	13	11	13	14	14	14	15
Propranolol	10	10	10	10	9	8	10	9	9	10
Diastolic blood pressure										
Bendrofluazide	5	6	6	6	6	6	6	7	7	6
Propranolol	4	5	6	5	6	4	5	5	5	4

The five and a half year cumulative percentages of people lapsing from follow up (Fig. 3) were about 19% and include losses of about 3.5% due to participants moving house.

The total five and a half year cumulative percentages of men who stopped taking their randomized treatment, including both those withdrawn from their randomly allocated regimen but continuing on follow up and those lapsing from the trial, were 43% of the bendrofluazide group, 42% of the propranolol group, and 47% of the placebo group. For women the figures were 33%, 40%, and 40% respectively. The cumulative percentages of people not taking either primary active drug by five and a half years were smaller: 33% of men originally randomized to bendrofluazide and 34% of men randomized to propranolol and 28% and 31% respectively of women.

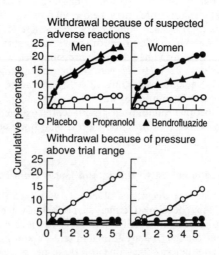

Figure 2. *Cumulative percentages withdrawn from randomized treatment.*

Table V *Principal reasons for withdrawal from randomized treatment. Numbers of reports and rates/1000 patient years†*

	Men						Women					
	Bendrofluazide		Propranolol		Placebos		Bendrofluazide		Propranolol		Placebos	
	No	Rate	No	Rate	No	Rate	No	Rate	No	Rate	No	Rate
Impaired glucose tolerance	60	7.7***	27	3.4	53	3.3	46	5.9***	16	2.1	31	2.0
Gout‡	100	12.8***	12	1.5	14	0.9	12	1.5***	0		0	
Impotence	98	12.6***	50	6.3***	20	1.3	0		0		0	
Raynaud's phenomenon	0		41	5.1***	3	0.2	2	0.3	34	4.5***	4	0.3
Skin disorder	6	0.8	12	1.5**	5	0.3	3	0.4	9	1.2**	2	0.1
Dyspnoea	1	0.1	57	7.1***	7	0.4	2	0.3	53	7.1***	3	0.2
Lethargy	28	3.6***	42	5.3***	8	0.5	13	1.7***	62	8.3***	4	0.3
Nausea, dizziness, or headache	33	4.2***	33	4.1***	22	1.4	58	7.4***	70	9.4***	27	1.8
Pressure at or above levels requiring change of treatment	8	1.0***	33	4.1***	611	38.2	11	1.4***	24	3.2***	400	26.0

†Patient years of observation relates here only to years accrued before withdrawal of randomized treatment.
‡Defined as symptoms plus serum urate values in excess of 500 μmol/l in men, 450 μmol/l in women.
** $p < 0.01$; *** $p < 0.001$; p values are for comparison of rate on individual active drug with rate on placebos.

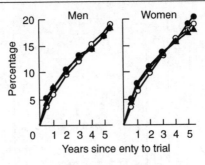

Years since enty to trial

o Placebo   ● Propranolol   ▲ Bendrofluazide

Figure 3. *Cumulative rates of people lapsing from follow up.*

Primary results

The principal findings, directly answering those questions specified at the design stage, were as follows (table VI, Fig. 4).

Stroke—The event rate of stroke was significantly reduced in people randomized to receive active treatment. There were 60 strokes in the actively treated group and 109 in the placebo group (p = 0.0006 on once off testing, p < 0.01 on sequential analysis). The percentage difference between the rates in the treated and placebo groups (1.4 and 2.6 per 1000 patient years respectively) was 45%. The absolute difference was 1.2 strokes per 1000 patient years.

Coronary events—The overall coronary event rate was not significantly affected by treatment (222 in the treated group, 234 in the placebo group, with rates per 1000 patient years of 5.2 and 5.5 respectively).

All cardiovascular events—The cardiovascular event rate was significantly reduced in the actively treated group. There were 286 such events on active treatment and 352 in the placebo group (p = 0.01 on once off testing, p < 0.05 on sequential analysis). Rates per 1000 patient years were 6.7 and 8.2 respectively, with a percentage difference of 19% between rates for treated and placebo groups and an absolute differcnce of 1.6 events per 1000 patient years. Results in this category are dominated by figures for coronary events, which considerably exceeded the numbers of strokes.

Analysis also showed that the all cause mortality was almost identical in the two groups. There were 248 deaths in the treated group and 253 in those taking placebo tablets, giving rates of 5.8 and 5.9 per 1000 patient years respectively.

Subgroups

Individual active drug (tables VII and VIII)

Stroke—Both drugs were associated with reduced stroke rates. Eighteen strokes occurred in the bendrofluazide group, 42 on propranolol, and 109 on placebo (rates of 0.8, 1.9, and 2.6 per 1000 patient years respectively). The percentage reduction on bendrofluazide was significantly (p = 0.002) greater than that on propranolol.

Table VI *Main events for both sexes together. Numbers and rates per 1000 patient years*

	Active treatment†		Placebos		% Difference‡ (95% confidence limits)	Absolute difference/ 1000 patient years§ (95% confidence limits)
	No	Rate	No	Rate		
Strokes						
Fatal	18	0.4	27	0.6	34	0.2
Non-fatal	42	1.0	82	1.9	49	0.9
Total	60	1.4	109	2.6	45 (25, 60)	1.2 (0.6, 1.7)
Coronary events						
Fatal	106	2.5	97	2.3	−9	−0.2
Non-fatal	116	2.7	137	3.2	16	0.5
Total	222	5.2	234	5.5	6 (−13, 21)	0.3 (−0.7, 1.3)
All cardiovascular events*	286	6.7	352	8.2	19 (5, 31)	1.6 (0.4, 2.7)
All cardiovascular deaths	134	3.1	139	3.3	4 (−22, 24)	0.1 (−0.6, 0.9)
Non-cardiovascular deaths	114	2.7	114	2.7	0 (−29, 23)	0.0 (−0.7, 0.7)
All deaths	248	5.8	253	5.9	2 (−16, 18)	0.1 (−0.9, 1.2)

*Not necessarily equal to the total of strokes plus coronary events because it also includes 'other relevant deaths' and death due to other cardiovascular causes such as ruptured aneurysms.

†Randomized either to bendrofluazide or to propranolol.

‡Percentage difference between rates on active and on placebo therapy.

§Absolute difference between rates on active treatment and on placebo therapy.

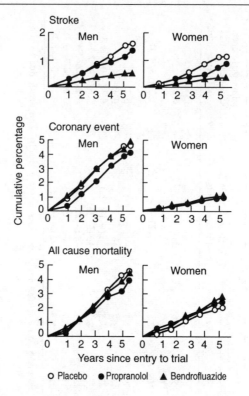

Figure 4. *Cumulative percentages of people with terminating events (stroke, coronary events, and all cause mortality) by sex and by randomized treatment.*

All cardiovascular events — Both drugs were associated with slightly reduced rates (6.6 per 1000 patient years on bendrofluazide, 6.7 on propranolol, and 8.2 on placebo). There was no significant difference between the effects of the individual active drugs (p = 0.76).

For *coronary events* and for *all cause mortality* there were no statistically significant differences between the effect associated with bendrofluazide and those associated with propranolol (p = 0.24 and 0.71 respectively).

Sex (tables VII and IX)

Strokes, coronary events, and all cardiovascular events — There were no statistically significant differences between men and women in their percentage response to active treatment (p values 0.63, 0.45, and 0.62 respectively).

All cause mortality — There was a benefit associated with treatment in men (157 deaths on active treatment and 181 deaths on placebo (7.1 and 8.2 per 1000 patient years respectively)) but the opposite effect in women (91 deaths in the treated group compared with 72 on placebo, giving rates of 4.4 and 3.5 per 1000

Table VII *Numbers, person years of observation, and principal events by sex and drug regimen at randomization. Rate – rate/1000 patient years*

	Men								Women							
	Bendrofluazide		Propranolol		Active treatment		Placebos		Bendrofluazide		Propranolol		Active treatment		Placebos	
No entered:	2 238		2 285		4 523		4 525		2 059		2 118		4 177		4 129	
Patient years:	10 945		11 184		22 129		22 190		10 274		10 508		20 782		22 471	
	No	Rate	No	Rate	No	Rate	No	Rate	No	Rate	No	Rate	No	Rate	No	Rate
Stroke																
Fatal	0	0.0	6	0.5	6	0.3	13	0.6	4	0.4	8	0.8	12	0.6	14	0.7
Non-fatal	11	1.0	20	1.8	31	1.4	52	2.3	3	0.3	8	0.8	11	0.5	30	1.5
Total	11	1.0	26	2.3	37	1.7	65	2.9	7	0.7	16	1.5	23	1.1	44	2.1
Coronary events																
Fatal	50	4.6	38	3.4	88	4.0	87	3.9	9	0.9	9	0.9	18	0.9	10	0.5
Non-fatal	49	4.5	47	4.2	96	4.3	113	5.1	11	1.1	9	0.9	20	1.0	24	1.2
Total	99	9.0	85	7.6	184	8.3	200	9.0	20	1.9	18	1.7	38	1.8	34	1.7
All cardiovascular events	113	10.3	112	10.0	225	10.2	272	12.3	27	2.6	34	3.2	61	2.9	80	3.9
All cardiovascular deaths	56	5.1	48	4.3	104	4.7	112	5.1	13	1.3	17	1.6	30	1.4	27	1.3
Non-cardiovascular deaths	26	2.4	27	2.4	53	2.4	69	3.1	33	3.2	28	2.7	61	2.9	45	2.2
All deaths	82	7.5	75	6.7	157	7.1	181	8.2	46	4.5	45	4.3	91	4.4	72	3.5

Table VIII *Principal events by randomly allocated drug, both sexes together**

	Bendrofluazide		Propranolol		Placebos		% Difference		Absolute difference/1000 patient years	
	No	Rate	No	Rate	No	Rate	Bendrofluazide	Propranolol	Bendrofluazide	Propranolol
Strokes	18	0.8	42	1.9	109	2.6	67	24	1.7	0.6
Coronary events	119	5.6	103	4.8	234	5.5	−2	13	−0.1	0.7
All cardiovascular events	140	6.6	146	6.7	352	8.2	20	18	1.7	1.5
Non-cardiovascular deaths	59	2.8	55	2.5	114	2.7	−4	5	−0.1	0.1
All deaths	128	6.0	120	5.5	253	5.9	−2	6	−0.1	0.4

* Apparent discrepancies are due to rounding in the figures presented for the rates.

Table IX *Principal events by sex*

	Active treatment		Placebo treatment		% Difference	Absolute difference/1000 patient year
	No	Rate	No	Rate		
Men						
Strokes	37	1.7	65	2.9	43	1.3
Coronary events	184	8.3	200	9.0	8	0.7
All cardiovascular events	225	10.2	272	12.3	17	2.1
Non-cardiovascular deaths	53	2.4	69	3.1	23	0.7
All deaths	157	7.1	181	8.2	13	1.1
Women						
Strokes	23	1.1	44	2.1	48	1.0
Coronary events	38	1.8	34	1.7	−11	−0.2
All cardiovascular events	61	2.9	80	3.9	25	1.0
Non-cardiovascular deaths	61	2.9	45	2.2	−34	−0.8
All deaths	91	4.4	72	3.5	−25	−0.9

Table X. *Principal events by entry smoking habit, both sexes together*.*

| | Bendrofluazide | | | | Propranolol | | | | Placebos | | | | % Benefit | | | | Absolute benefit/1000 patient years | | | |
| | Smokers 6206 | | Non-smokers 14 913 | | Smokers 6 056 | | Non-smokers 15 498 | | Smokers 12 352 | | Non-smokers 30 152 | | Bendrofluazide v placebos | | Propranolol v placebos | | Bendrofluazide v placebos | | Propranolol v placebos | |
Patients years:	No	Rate	No	Rate	No	Rate	No	Rate	No	Rate	No	Rate	Smokers	Non-smokers	Smokers	Non-smokers	Smokers	Non-smokers	Smokers	Non-smokers
Strokes	6	1.0	12	0.8	26	4.3	16	1.0	48	4.0	60	1.9	75	59	−8	47	3.1	1.1	−0.3	0.9
Coronary events	57	9.3	62	4.1	57	9.5	45	2.9	102	8.5	131	4.3	−9	4	−12	33	−0.8	0.2	−1.0	1.4
All cardiovascular events	65	10.6	75	5.0	84	14.0	61	3.9	157	13.2	193	6.3	20	21	−7	38	2.6	1.3	−0.9	2.4
Non-cardiovascular deaths	17	2.8	40	2.7	19	3.2	36	2.3	51	4.3	63	2.1	35	−30	26	−12	1.5	−0.6	1.1	−0.3
All deaths	47	7.7	79	5.3	54	9.1	66	4.2	119	10.1	134	4.4	23	−20	10	3	2.4	−0.9	1.0	0.1

*Number of events do not always tally with those in other tables, because there were 76 people for whom the smoking habit at entry was not recorded and who are not included here.

patient years respectively). The difference between the sexes was significant (p = 0.05).

Cigarette smoking (table X, Figs 5 and 6)

Effects of active treatment — When the percentage response to active treatment in non-smokers was compared with that in smokers, no statistically significant differences were found for stroke (p = 0.19), for coronary events (p = 0.08), or

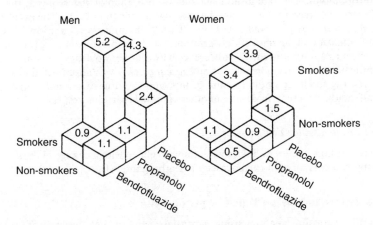

Figure 5. *Incidence of stroke per 1000 person years of observation according to randomized treatment regimen and cigarette smoking status at entry to trial.*

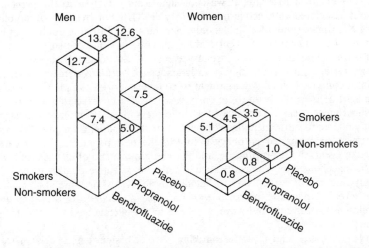

Figure 6. *Incidence of coronary events per 1000 person years of observation according to randomized treatment regimen and cigarette smoking status at entry to trial.*

for all cause mortality (p = 0.19). For all cardiovascular events the difference between smokers and non-smokers was just statistically significant (p = 0.05), with the greater benefit in non-smokers.

Comparison of drugs—Smoking habit did not affect the response to bendrofluazide but was important when the response to propranolol was considered. For *stroke* the event rate was reduced in smokers and in non-smokers on bendrofluazide but only in non-smokers on propranolol and the difference between the two drugs was significant in this respect (p = 0.03). For *coronary* events the rate was not affected in smokers or non-smokers on bendrofluazide or in smokers on propranolol. The rate in non-smokers on propranolol was reduced, but the difference between the two drugs was not significant (p = 0.11). The event rate for *all cardiovascular events* was not affected in smokers or non-smokers on bendrofluazide or in smokers on propranolol. In non-smokers on propranolol the event rate was reduced. The difference between the two drugs was statistically significant (p = 0.01). The rate for *all cause mortality* was not affected by either active drug in either smoking habit group, and the drugs were not statistically significantly different from one another (p = 0.11) in this respect.

Non-cardiovascular deaths

There was no evidence to link any treatment regimen with a change in death rates in any category of non-cardiovascular deaths (table XI).

Risk factors and the prediction of personal risk

Logistic regression analyses using entry data showed that the level of systolic blood pressure at entry was significantly associated with the risk of stroke, coronary events, all cardiovascular events, and all cause mortality.

Although the level of entry systolic pressure was significantly associated with the subsequent stroke rate, it was not significantly related to the percentage benefit associated with active treatment (table XII). For the sake of simplicity table XII shows results for stroke only, but the general finding applies similarly to the other categories of events: although entry systolic pressure was a risk factor for the development of coronary events, all cardiovascular events, and all cause mortality, it was not significantly associated with any percentage changes in rates conferred by active treatment in comparison with the placebo regimen. The level of diastolic pressure at entry was less clearly associated with the risk of subsequent events. The level of systolic pressure at six months was significantly related to subsequent development of stroke in patients on any of the three treatment regimens, of coronary events in the placebo group, of all cardiovascular events in the propranolol and placebo groups, and of all cause mortality in the placebo group. The percentage benefit due to active treatment, however, in any category of events, was not related to the level of pressure at six months. The level of diastolic pressure was less clearly related to the subsequent development of events and was not related to the percentage benefit associated with treatment.

Age, male sex, and cigarette smoking were also significantly related to the subsequent development of stroke, coronary events, all cardiovascular events, and all cause mortality (table XIII). The presence of an ischaemic pattern on

Table XI *Non-cardiovascular deaths*

	Men								Women							
	Bendrofluazide		Propranolol		Active treatment		Placebos		Bendrofluazide		Propranolol		Active treatment		Placebos	
	No	Rate	No	Rate	No	Rate	No	Rate	No	Rate	No	Rate	No	Rate	No	Rate
Malignant*	17	1.6	18	1.6	35	1.6	54	2.4	24	2.3	23	2.2	47	2.3	36	1.8
Infection	2		0		2		4		0		0		0		1	
Pancreatitis	1		0		1		0		1		1		2		2	
Accident	2		5		7		3		1		1		2		1	
Suicide	2		0		2		1		3		0		3		0	
Other/not known	2		4		6		7		4		3		7		5	
Total	26	2.4	27	2.4	53	2.4	69	3.1	33	3.2	28	2.7	61	2.9	45	2.2

* Malignant deaths were analysed by site; there was no evidence that any treatment regimen was associated with an excess of tumours at any site.

Table XII *Stroke rates and percentage benefit associated with treatment by (a) entry systolic pressure and (b) entry diastolic pressure, both sexes together*

	Entry systolic pressure (mm Hg)					Entry diastolic pressure (mm Hg)			
	<150	150–159	160–169	170–179	≥180	<95	95–99	100–104	≥105
Strokes/1000 patient years									
Bendrofluazide	0.6	1.0	0.4	0.8	1.2	0.2	1.3	1.0	1.1
Propranolol	1.2	1.0	2.9	2.2	2.3	1.5	2.4	2.2	1.6
Active treatment	0.9	1.0	1.7	1.5	1.7	0.8	1.8	1.6	1.4
Placebos	1.4	1.8	2.5	3.2	3.9	1.4	2.3	3.1	4.4
% Benefit									
Bendrofluazide	60	46	83	75	70	88	44	68	75
Propranolol	19	46	−16	29	42	−9	−6	31	65
Active treatment	39	46	33	52	56	40	18	50	69

Table XIII Contribution of baseline variables to risks of developing a terminating event in the trial.

	Systolic blood pressure— 10 mm Hg increase	Diastolic blood pressure— 4 mm Hg increase	Age— 5 years increase	Women: men	Smokers: non-smokers	Ischaemic ECG: non-ischaemic ECG*	Cholesterol 6.5: 6.5 mmol/l	Quetelet index— 3.0 kg/m increase	Placebos: active	Propranolol: bendrofluazide
Stroke										
p value	0.02	0.003	$<10^{-4}$	0.002	$<10^{-4}$	0.06	0.43	0.19	0.0006	0.002
RR (95% CL)	1.14	1.21	1.46	0.60	2.29	1.65	1.14	0.91	1.74	2.30
	(1.02–1.28)	(1.07–1.37)	(1.28–1.66)	(0.45–0.83)	(1.68–3.13)	(1.00–2.69)	(0.83–1.56)	(0.80–1.05)	(1.26–2.40)	(1.33–3.99)
Coronary events										
p value	0.001	0.71	$<10^{-4}$	$<10^{-4}$	$<10^{-4}$	$<10^{-4}$	$<10^{-4}$	0.008	0.60	0.24
RR (95% CL)	1.10	1.02	1.32	0.15	2.27	2.13	1.76	1.13	1.05	0.85
	(1.03–1.18)	(0.94–1.10)	(1.23–1.43)	(0.11–0.19)	(1.87–2.76)	(1.56–2.90)	(1.45–2.14)	(1.03–1.23)	(0.87–1.27)	(0.64–1.11)
Cardiovascular events										
p value	0.001	0.05	$<10^{-4}$	$<10^{-4}$	$<10^{-4}$	$<10^{-4}$	$<10^{-4}$	0.19	0.01	0.76
RR (95% CL)	1.11	1.07	1.38	0.23	2.36	1.94	1.60	1.05	1.23	1.04
	(1.04–1.17)	(1.00–1.14)	(1.29–1.47)	(0.19–0.28)	(2.01–2.79)	(1.48–2.54)	(1.35–1.89)	(0.98–1.13)	(1.05–1.46)	(0.81–1.32)
All deaths										
p value	0.03	0.82	$<10^{-4}$	$<10^{-4}$	$<10^{-4}$	$<10^{-4}$	0.005	0.78	0.86	0.71
RR (95% CL)	1.07	1.01	1.41	0.41	1.99	2.27	1.31	1.01	1.02	0.95
	(1.01–1.14)	(0.94–1.09)	(1.31–1.52)	(0.34–0.50)	(1.66–2.40)	(1.72–3.00)	(1.09–1.58)	(0.93–1.10)	(0.85–1.22)	(0.73–1.23)

RR — relative risk; CL — confidence limits;
*1, 4, 5 (one or more code present).

the electrocardiogram and a high serum cholesterol concentration were risk factors for coronary events, all cardiovascular events, and all cause mortality. Quetelet's body mass index was a risk factor for coronary events only.

The value of these logistic regressions, using all entry data including blood pressure, in identifying those individuals who would suffer any event was then assessed, using data for the placebo group. Risk scores were calculated, based on multiple logistic regressions; there was considerable overlap between scores for event and non-event groups. If 80% of people in the (relatively small) group who experienced a coronary event were to be correctly identified using entry data at least 38% of the (much larger) group who did not have a coronary event would have been incorrectly classified. Discrimination of people likely to develop stroke was even less precise, so that overall there is no method which would enable a doctor to give a useful prediction to an individual patient.

DISCUSSION

In answer to the principal questions specified at the design stage these results provide clear evidence that active treatment was associated with a reduction in stroke rate in this mildly hypertensive population and show no clear overall effect on the incidence of coronary events. Active treatment had no evident effect on the overall all cause mortality, but there was a beneficial effect in men and an adverse effect in women. With this exception, there is no clear evidence that the effects of treatment differed in the two sexes.

The reduction in the stroke rate attributable to anti-hypertensive treatment both confirms and adds to the results of earlier national studies.[18] In this predominantly white population the reduction was shown in both sexes. Since the percentage reduction was not related to pressure level at entry, the finding seems to apply equally to the complete range of pressure studied. The result is, of course, specific to the trial population. Extrapolation from these results to pressures outside the range, or to older people, would need further evidence.

This reduction in stroke rate due to active treatment first became evident in 1983 (p < 0.01 on the sequential analysis), but the reduction in stroke rate had to be balanced against possible adverse effects of active treatment on other event rates, and since the overall mortality rates were no different in treated and placebo groups it was thought impossible to use the results as a basis for definitive recommendations about the management of mild hypertension, and the study was continued.

Comparison of the results for the individual active drugs is inextricably linked with what is perhaps the most interesting aspect of these analyses, the difference between results for non-smokers and for smokers.

Bendrofluazide, which reduced stroke rates in both non-smokers and smokers, was associated with a greater reduction in the stroke rate than was propranolol, which reduced the stroke rate in non-smokers only. Bendrofluazide was not associated with a reduction in the coronary event rate either in non-smokers or in smokers, but propranolol reduced this rate in non-smokers. Blood pressure control was also better in the bendrofluazide group (Fig. 1m, Table IIm). The dose of bendrofluazide was fixed and, if this did not reduce pressure to target level within a specified period, supplementary therapy was immediately introduced. The results suggest that, for the groups overall, the simple regimen of a fixed dose proceeding automatically to the addition of supplementary therapy if necessary may have been more effective in achieving target pressure than the more complicated propranolol regimen, which necessitated increasing the dose before adding a supplementary drug. The diminished antihypertensive efficacy of propranolol in smokers when compared with non-smokers[19] presumably also contributed to this difference in pressure control.

The total picture, then, is that the incidence of stroke was reduced on active treatment with either drug but, whereas bendrofluazide was equally effective in smokers and in non-smokers, propranolol seemed to be relatively ineffective in smokers. For coronary events rates were unaffected by treatment overall, but within subgroups propranolol apparently reduced the coronary event rate in non-smokers though not in smokers. At one stage during the earlier part of the trial there was a trend towards an excess of fatal coronary events in men randomized to bendrofluazide, and concern about this suggestion of a serious drug adverse effect prompted the setting up of a substudy of the relationship between bendrofluazide and ventricular ectopic beats.[20] The numbers of events were small and no firm evidence of an association between bendrofluazide and coronary death has been established. The Multiple Risk Factor Intervention Trial Research Group also referred to the possibility, suggested by a subgroup analysis, that hypertensive men who had abnormal baseline electrocardiograms and were randomized to the active intervention group experienced an excess of fatal coronary events.[21] The authors did warn, however, that these data should not be overemphasised, and a statistical discussion of subgroup analyses subsequently stated that there were inadequate grounds for supposing that the intervention group had been harmed by the active regimen.[14] In the MRC trial the non-cardiovascular causes of death give no evidence that either drug altered the incidence of carcinoma, but numbers are small and the time which has elapsed since patients first took the drug is short. No other non-cardiovascular cause of death is clearly associated with either active drug. The overall all cause mortality rate was unaffected by treatment.

About one eighth of those randomized to the placebo group needed active treatment because their blood pressures rose above the limit considered permissible. Higher associated levels of morbidity and mortality might have been expected had the pressure in these people been allowed to rise further, but, because they form a selected group and because the effects of randomization have been lost, it is not possible to arrive at valid comparative figures for event rates in this group.

There have been two other large trials which have a considerable bearing on the treatment of mild hypertension. The first is the Australian National Blood Pressure Study,[22] which has the greatest similarities with the MRC trial, since it was based on the comparison of actively treated and placebo groups of subjects, and the second is the US Hypertension Detection and Follow-up Program,[23] which was unlike the other two trials in that it compared a group of subjects treated with the greatest care to ensure compliance in a hospital clinic (stepped care) with a comparable group referred back to their own physicians for treatment (referred care).

In its design, therefore, the Australian trial is the one that first needs comparative assessment. The numbers included were much smaller and the decision to stop the trial was made after about 14 000 patient years of observation, which may be contrasted with the nearly 90 000 patient years of observation in the MRC trial. This is in part because the Australian trial was stopped when the results were just of marginal statistical significance, so that the number of morbid events was quite small. Nevertheless, the main conclusion was that there was a reduction in the incidence of stroke in the treated group.

The Hypertension Detection and Follow up Program trial cannot be directly compared with either of these two trials, since it had a completely different design and compared one form of treatment with another without a placebo group. Its final conclusions were that in the more intensively treated group (stepped care) there was a reduction in both cardiovascular and non-cardiovascular mortality. This latter finding is different from that of the Australian and MRC trials. A further obvious difference is that the populations studied were dissimilar, not only in the number of blacks in the US trial but in the degree of cardiovascular morbidity in the populations studied. This is brought out by comparison of the mortality rates in the various trials. The Australian and MRC trials are very alike, but mortality from stroke was nearly three times greater among the appropriate part (stratum I) of the referred care group of the US trial than among the placebo group of the MRC trial, mortality from coronary heart disease over two times greater, and all cause mortality also nearly three times greater.[18] Various aspects of the US trial have received comment[24,25]; because of the quite different aims of the MRC and the US trials, and because they involved quite different types of medical care, it would be inappropriate to extrapolate from the Hypertension Detection and Follow up Program in considering what advice should be given to patients with mild hypertension in Britain.

Can advice be based on conclusions drawn from the present MRC trial?

CONCLUSIONS

The trial has shown that if 850 mildly hypertensive patients are given active antihypertensive drugs for one year about one stroke will be prevented. This is an important but an infrequent benefit. Its achievement subjected a substantial percentage of the patients to chronic side effects, mostly but not all minor. Treatment did not appear to save lives or substantially alter the overall risk of coronary heart disease. More than 95% of the control patients remained free of any cardiovascular event during the trial

Neither of the two drug regimens had any clear overall advantage over the other. The diuretic was perhaps better than the β blocker in preventing stroke, but the β blocker may have prevented coronary events in non-smokers.

For all categories of events, and in both treated and placebo groups, rates were lower in non-smokers than in smokers, adding to previous evidence that starting smoking considerably increases the risk of cardiovascular disease. For stroke and also for all cardiovascular events the difference between rates in smokers and non-smokers was greater than the effect of drug treatment.

ACKNOWLEDGEMENTS

The working party thanks the general practitioners and nurses who joined the research framework and without whose efforts the trial would have been impossible; this framework was established by Dr W. E. Miall with the help of Mrs G. R. Barnes and maintained from 1983 onwards by Dr G. Greenberg with the help of Mrs C. W. Browne; Professor H. D. Tunstall Pedoe for his work in arbitrating on the assessment of all terminating events; Professor T. P. Whitehead and Mr P. M. G. Broughton and the staff of the Wolfson Research Laboratories, Queen Elizabeth Medical Centre, Birmingham, for carrying out the biochemical analyses; Duncan, Flockhart and Co Ltd for tablets of bendrofluazide and placebo; Imperial Chemical Industries Ltd for financial support and for tablets of propranolol and placebo; CIBA Laboratories for supplies of guanethidine; and Merck Sharp and Dohme Ltd for a mobile screening unit, funds for its staffing, and supplies of methyldopa.

REFERENCES

1. MRC Working Party on Mild to Moderate Hypertension. Randomised controlled trial of treatment for mild hypertension: design and pilot trial. *Br Med J* 1977; ii: 1437–40.
2. Hamilton, M., Thompson, E. N., Wisniewski, T. K. M. The role of blood pressure control in preventing complications of hypertension. *Lancet* 1964; i: 235–8.
3. Veterans Administration Co-operative Study Group. Effects of treatment on morbidity in hypertension. Results in patient, with diastolic blood pressures averaging 115 through 129 mm Hg. *JAMA* 1967; **202:** 1028–34.
4. Carter, A. B. Hypotensive therapy in stroke survivors. *Lancet* 1970; i: 485–9.
5. Veterans Administration Co-operative Study Group. Effects of treatment on morbidity in hypertension II. Results in patients with diastolic blood pressure averaging 90 through 114 mm Hg. *JAMA* 1970; **213:** 1143–52.
6. Miall, W. E., Chinn, S. Screening for hypertension: some epidemiological observations. *Br Med J* 1974; iii: 595–600.
7. Registrar General. *Statistical Review of England and Wales, 1970.* London: HMSO, 1971.
8. Miall, W. E., Brennan, P. J. Observations on the natural history of mild hypertension in the control groups of therapeutic trials. In: Gross, F., Strasser, T., eds. *Mild hypertension, natural history and management.* Tunbridge Wells: Pitman Medical, 1979: 38–43.
9. Wright, B. M., Dore, C. F. A random-zero sphygmomanometer. *Lancet* 1970; i: 337–8.
10. Rose, G. A., Holland, W. W., Crowley, E. A. A sphygmomanometer for epidemiologists. *Lancet* 1964; i: 296–300.
11. Aho, H., Harmsen, P., Hatano, S., Marquardsen, W., Smirnov, V. E, Strasser, T. Cerebrovascular disease in the community; results of a WHO collaborative study. *Bull WHO* 1980; **58:** .113–30.
12. World Health Organisation Regional Office for Europe. *Myocardial infarction community registers.* Copenhagen: WHO, 1976. (Public Health in Europe No 5.)
13. Pocock, S. J. Group sequential methods in the design and analysis of clinical trials. *Biometrika* 1977; **64:** 191–8.
14. Pocock, S. J. Current issues in the design and interpretation of clinical trials. *Br Med J* 1985; **290:** 39–42.

15. Baker, R. J., Nelder, J. A. *The GLIM system*. Release 3. Oxford: Numerical Algorithms Group.

16. Brennan, P. J., Greenberg, G., Miall, W. E., Thompson, S. G. Seasonal variation in arterial blood pressure. *Br Med J* 1982; **285**: 919–23.

17. Medical Research Council Working Party on Mild to Moderate Hypertension. Adverse reactions to bendrofluazide and propranolol for the treatment of mild hypertension. *Lancet* 1981; ii: 539–43.

18. WHO/ISH Mild Hypertension Liaison Committee. Trials of the treatment of mild hypertension. An interim analysis. *Lancet* 1982; i: 149–56.

19. Greenberg, G., Thompson, S. G. The effect of smoking on blood pressure in mild hypertensives, and on their response to antihypertensive treatment. *Journal of Hypertension* 1984; **2**: 553.

20. Medical Research Council Working Party on Mild to Moderate Hypertension. Ventricular extrasystoles during thiazide treatment: sub-study of MRC mild hypertension trial. *Br Med J* 1983; **287**: 1249–53.

21. Multiple Risk Factor Intervention Trial Research Group. Multiple Risk Factor Intervention Trial. Risk Factor Changes and Mortality Results. *JAMA* 1982; **248**: 1465–77.

22. Australian National Blood Pressure Study Management Committee. The Australian therapeutic trial in mild hypertension. *Lancet* 1980; i: 1261–7.

23. Hypertension Detection and Follow-up Program Co-operative Group. Five-year findings of the hypertension detection and follow-up program. (1) Reduction in mortality of persons with high blood pressure, including mild hypertension (2) Mortality by race/sex and age. *JAMA* 1979; **242**: 2562–77.

24. Alderman, M. H., Madhavan, S. Management of the hypertensive patient: a continuing dilemma. *Hypertension* 1981; **3**: 192–7.

25. Toth, P. J., Horwitz, R. I. Conflicting clinical trials and the uncertainty of treating mild hypertension. *Am J Med* 1983; **75**: 482–8.

10 Testing a test: three critical steps

David Mant

INTRODUCTION

The paper analysed in this chapter (Dobbs and Fleming 1987) was written by a general practice trainee. The research project which it describes involved the co-operation of ten general practitioners, three trainees, 521 patients, and a local hospital laboratory. To organize the study and to analyse and publish the results was a great achievement. The purpose of this book is to encourage critical reading, and in some ways I would rather not be critical of this paper. Indeed I have selected it for its strengths. Very little general practice research deals so clearly with the diagnostic process. The paper is also well written and accessible – a major advantage from a teaching point of view.

The paper concerns the diagnosis of urinary tract infection using a symptom score. The principles which apply in its appraisal would apply equally to other papers about diagnostic tests. The three critical steps in dealing with any paper assessing the diagnostic process are:

(1) identify the key clinical questions;

(2) assess the gold standard;

(3) check the 4 box analysis.

The first step – identifying the key clinical question – is common to all critical reading.

STEP 1: KEY CLINICAL ISSUE

The key clinical issue for me is whether I should apply the symptom score in clinical practice? If I make the effort to calculate a score when I suspect urinary tract infection, will it be of any great benefit to my patients? The first step is therefore to scan the paper quickly to answer this question.

When I first scanned the paper I picked out two tables and one figure which looked clinically interesting. The first was Table 3 which gives the 'SHU scores' for the various symptoms and dipstick tests examined. It is necessary to look at the appendix to work out what the SHU score actually is, but some of you will recognize the similarity between the score and the likelihood ratio and will realize that it allows you to make a very rapid assessment of clinical value. Even if you don't know about likelihood ratios it is clear that the only symptoms which score highly (± 3 or more) in children are presence or absence of dysuria. In adults, absence of frequency is helpful at all ages, but presence of haematuria and absence of nocturia are age dependent. My response was to make a mental note that information about the presence or absence of symptoms is not necessarily of equal value, but to try to remember to ask about these symptoms in the consultation. I also noted that symptoms are much less helpful than dipstick testing, particularly with a nitrite stick.

My attention was next drawn to Table 4, and in the margin I drew the following table:-

	Using score	Not Using Score
Not infected but treated	25%	38%
Not treated but infected	15%	28%

This struck me as interesting, as it suggested that by using the score I could almost halve the number of patients with UTIs which I failed to treat (at least until the MSU result was available) and could also decrease by a third the number of patients treated unnecessarily.

Finally, I looked at Figure 1. This has the potential to answer the question of whether it is better to elicit symptoms, or simply to listen. It is a little confusing because (as we shall see) the vertical axes are mislabelled; the results reported refer to the sensitivity and specificity rather than the positive and negative predictive values. However, again it is obvious from the figure that it makes little difference whether you elicit or listen for symptoms of dysuria, haematuria, or offensive urine. On the other hand it matters more for nocturia and frequency.

So at the first glance the findings seem clinically relevant and interesting and to merit closer examination to see if they should be believed. To answer this question of believability it is necessary to proceed to Steps 2 and 3, which take longer to sort out.

STEP 2: GOLD STANDARD

The validity of the whole paper depends on the adequacy of the 'gold standard' for urinary tract infection against which the SHU items are assessed. It is crucial that the cases of urinary tract infection are diagnosed with reasonable certainty according to best current practice. The authors discuss the difficulty of establishing a gold standard in their introduction. It is clear that there is debate about the validity of using a colony count of 100 thousand organisms per ml as a criterion for infection but there is little practical alternative and my main anxieties lie elsewhere. First, the authors were forced to add the additional gold standard criterion of positivity (at the laboratory) for blood, protein, or nitrite because of local laboratory practice. Although a reference is cited which gives the negative predictive value of such testing as 96 per cent (i.e. only four out of 100 infected urines will not be cultured if this criterion is applied first) the use of this additional criterion is unacceptable when one objective of the study was to assess dipstick testing (as part of the SHU score) in the practice. Secondly, because of the decision to rely on normal clinical practice in specimen collection (how well do you ensure MSUs are mid-stream and the perineum has been cleaned before collection?) and reliance on the routine laboratory collecting system (how long can some specimens take to reach the laboratory in your practice?) there must be anxieties about the 'gold standard'. About 10 percent of patients were excluded because of culture of mixed growths or identification of bacterial inhibitors. Thirdly, the concordance data presented and the discussion of the' possibility of using the probability scores of urinary tract infection to differentiate significant infection from contamination, did not help me in the interpretation of the main results.

The required methodology is simple—a comparison needs to be made between the performance of the SHU score against the existing 'gold-standard'. In my view to try to go beyond this by suggesting that the new strategy could be used to define significant infection, is likely to confuse the reader and to introduce tautology.

STEP 3: FOUR BOX ANALYSIS

The conventional four box analysis, is widely accepted as the appropriate method for assessing symptoms and diagnostic tests and also fits best with clinical common sense. The third step is to apply it (correctly) to characterize a diagnostic test according to the parameters of sensitivity, specificity, and predictive value. The basic method for doing this is shown in Table 10.1. All the patients tested are put into one of four boxes, according to whether they have the disease or not (according to the gold standard) and whether or not the diagnostic test (or symptoms score) is positive or negative. For theoretical purposes the sensitivity and specificity are most useful because they are independent of the prevalence of the condition in the population in which the diagnostic test is used. This is because the sensitivity is calculated only in patients with the disease and the specificity

Table 10.1 *Four box analysis of frequency symptom (numbers estimated from percentages given in Table 1, Dobbs and Fleming 1987)*

Frequency symptom	Present	Proven infection Absent	Total	
Present	95	160	255	Positive predictive value 95/255 = 37%
Absent	20	130	150	Negative predictive value 130/150 = 87%
Total	115	290	405	
	Sensitivy 95/115 = 83%	Specificity 130/290 = 45%		

is calculated only in patients without the disease. The sensitivity provides a measure of the false negative results—how often the test misses people with disease. The specificity provides a measure of the false positive results—how often the test identifies patients as having the disease when they do not.

The sensitivity and specificity are important, but in order to assess precisely how a diagnostic test works in a practice population it is necessary to take a further step and to calculate the predictive value of a positive test and of a negative test. The predictive value of a positive test is the proportion of patients who test positive who actually have the disease; the predictive value of a negative test is the proportion of patients who test negative who are healthy. Predictive values vary in different clinical settings according to the prevalence of disease in that setting; positive predictive value is usually lower in general practice than in hospital. It is a great advantage of the paper reviewed here that the research is being done in general practice, so that the predictive value can be established directly rather than indirectly from hospital studies.

The results of this paper are not analysed by the traditional 'four box' method and this is obviously a major methodological weakness. Indeed, although predictive values are quoted they are incorrectly calculated. In the analysis section on page 101 the positive predictive value is defined as the percentage occurrence of symptoms in infected urines and the negative predictive value as 100 minus the occurrence in non-infected urines. As infected urine is the gold standard, these are in fact definitions of the sensitivity and specificity. The sensitivity of frequency as a symptom of infection is the proportion of people with infection who experience frequency (95/115 or 83 per cent). Conversely, the positive predictive value is the proportion of people with frequency who have a urinary tract infection (95/255 = 37 per cent). Therefore, an important step in interpreting this paper is to recalculate the data in Table 1 using the correct four box analysis.

This recalculation took me about 20 minutes with a pen and calculator, again estimating the numbers from the percentages given. The results are shown in Table 10.2. It can be seen at a glance that diagnosis of urinary tract infection is not easy. Of the various symptoms only haematuria gives a positive predictive value of more than 50 per cent. Although the surgery dipstick nitrite test gives a positive predictive value of 87 per cent, it must be remembered that a positive laboratory dipstick test was one of the criteria used for the gold standard. Similarly, the negative predictive values are all less than 90 per cent. This means that basing a diagnostic decision on any one individual symptom allows at least one in ten urinary tract infections to remain undetected and untreated. However, it is also clear that some symptoms are better than others at excluding disease—the absence of frequency being the best and the absence of nausea being the worst.

Table 10.2 *Sensitivity and predictive values (estimated from data given in Table 1, Dobbs and Fleming 1987)*

	Sensitivity (%)	Specificity (%)	Positive predictive value (%)	Negative predictive value(%)
Symptoms				
Frequency	83	45	37	87
Nocturia	64	64	42	82
Dysuria	70	60	41	84
Urgency	43	76	41	84
Haematuria	14	95	52	74
Offensive urine	22	89	44	74
Nausea	9	81	15	69
History				
Previous UTI	57	60	36	78
Short history	79	38	34	82
Previous IVP	15	94	50	74
Dipstick				
Protein	41	85	52	78
Blood	69	74	51	86
Nitrite	54	97	87	84

At this stage it is time to return to the SHU score. The method of generating the SHU score is shown in an appendix to the paper and it is not as complex as it first appears. Essentially, it is based on the likelihood ratio (sensitivity/1 − specificity) and a log value is taken so that the scores can be added together. As the scores are based on a relatively small number of patients, they may not be very accurate, but the application of the score is quite straightforward and could reasonably be used in general practice.

We looked at Table 4 in our first glance through the paper and noted the potential benefit of the SHU score compared with GP action. It is probably simplest to ignore the two middle columns of the table. The adequacy of the computer predicted probability model can only be examined by finding reference number 16 which in itself is rather sketchy. Similarly, the meaning of 'GP opinion' is unclear. In the method section, it states that GPs were asked to categorize the likelihood of infection as definite, probable, or maybe and the cut off point used is not stated. But we can use the four box approach to convert the data in columns 1 and 4 of Table 4 into predictive values. We can also use Figure 2 to look at the effect of using different SHU cut-off points. I did not draw attention to Figure 2 initially because the labelling of the horizontal axis is difficult to interpret but, with a bit of visual estimation, and a little more pencil and paper work, it does give the necessary additional data. My ten minute guestimates, by applying

Table 10.3 *Positive and negative predictive values according to different 'cut-offs' of SHU score for women aged 15–49 years*

SHU Score 'cut off'*	Positive predictive value (%)	Negative predictive value (%)
−4	43	92
−2	50	92
0	53	93
+2	66	90
+4	77	87
+6	81	86

*Score equal to or greater than cut-off value counting as positive test

the four box analysis to Figure 2 are shown in Table 10.3. These data allow the reader make his or her own choice of cut-off on the basis of the relative costs of false positive and false negative results in an individual clinical setting.

CONCLUSION

In the end, the proof of the paper is whether or not it allows you to decide to use the SHU score in your practice. The SHU scores derived in this paper cannot confidently be applied in other settings, because neither the gold standard nor the analysis were quite satisfactory. I already use nitrite dipsticks and my main interest is in the added value of the symptom score—which the paper doesn't really answer. But despite the criticisms made above, I would like to see the work described taken further because the paper did convince me that paying more formal attention to the predictive value of symptoms has the potential to improve the quality of my clinical care.

I think a more convincing study is worth doing and any reader who shares my interest might like to consider the following necessary ingredients of a follow-up:

1. The gold standard would be tighter. In particular, dipstick pretesting at the laboratory would be precluded if the value of dipsticks in the practice was to be as assessed.

2. The study would be bigger to increase the precision of the estimates made, particularly if separate scores were to be calculated for individual age–sex groups.

3. Formal measures of cost and benefit to the doctor and patient would be included in the study.

4. Routine general practitioner management would be more clearly defined. In this study the well-recognized variations in clinical management are not really discussed or taken into account.

It would be extremely interesting to have a further descriptive study along these lines, but it may be easier to obtain funding for a formal trial looking at resource costs and patient outcomes. This would allow better assessment of the impact of false negative results and also allow formal assessment of the savings in laboratory and prescribing costs.

Finally, anyone interested in following up issues of diagnostic test assessment and critical appraisal, and in particular the use of likelihood ratios in everyday practice, should read Sackett, Haynes, and Tugwell's *Clinical Epidemiology* (1991) — the classic textbook from McMaster University. This is the best book on clinical diagnosis I have read and is full of relevant clinical examples.

A simple scoring system for evaluating symptoms, history and urine dipstick testing in the diagnosis of urinary tract infection

F. F. Dobbs[1] and D. M. Fleming[2]

[1]GP Trainee, Birmingham and [2]Deputy Director, Royal College of General Practitioners, Birmingham Research Unit

Journal of the Royal College of GPs, **37**, 100–4 (1987)

Abstract

Patients presenting with symptoms suggestive of urinary tract infection were recruited in a general practice survey aimed at measuring the predictive value of symptoms, history and urine dipstick testing for diagnosing the presence of bacterial infection. Urine specimens were obtained from 87% of the 521 patients recruited. A diagnosis of infection was established by urine culture producing a colony count in a pure culture exceeding 100 000 organisms per ml or between 10 000 and 100 000 organisms per ml plus a minimum of 100 leucocytes per mm[3].

Occurrence rates for symptoms and other items of information in infected and non-infected groups were used to derive their positive and negative predictive values in making the diagnosis. The predictive value of volunteered symptoms was compared with that of elicited and volunteered symptoms combined. The positive predictive value of symptoms was increased where elicited symptoms were included but this was achieved at the cost of diminishing the negative predictive value. The occurrence rates were used to derive a mathematical model for diagnosing infection. The symptoms-history-urinalysis (SHU) score generated in this model compared well with a computer predicted probability. Both were substantially better than the assessment and action (decision to prescribe an antibiotic) of the recording doctor.

The scoring method described has been demonstrated in urinary tract infection but may be applied to any symptom combination related to a diagnosis for which there is an agreed definition.

INTRODUCTION

Studies of suspected urinary tract infection in general practice[1-5] indicate that in approximately 50% of possible cases a bacterial infection with a colony count exceeding 100 000 organisms per ml is not identified. However, a survey on urine culture by the Public Health Laboratory found that 94% of general practitioners start antibiotic treatment before receiving the culture report.[6] Thus many patients are prescribed antibiotics unnecessarily for this common complaint and perhaps many unnecessary and wasteful cultures of urine specimens are taking place. It may be difficult to differentiate the presence or absence of infection on clinical grounds alone,[1,7-9] though O'Dowd[5] believed it could be done if attention were paid to the psychosomatic features of the case.

Urine dipsticks have been advocated as an aid to diagnosis in the consulting room[10] and in recent years an increasing variety of combinations of dipsticks has become available. Zilva[11] reviewed their usefulness in routine screening of urine and drew attention to the correct storage procedures, the need for careful technique using freshly voided urine collected without contamination and the desirability of using the minimum combination of tests to achieve the desired objective. She considered that the urine must be examined by microscopy and microbiological methods if urinary tract infection is suspected. The relative cost of urine microscopy and culture (local private hospital cost £9.25) and dipstick (N-Labstix 10 p) must also be borne in mind.

The validity of the conventional criterion for a diagnosis of urinary tract infection (a colony count of 100 000 organisms per ml of urine) has been questioned by several authors.[3,4,12] Stamm[12] considered a level of 100 organisms per ml may be a sufficient criterion and stressed the need for microbiologists to be given adequate clinical information. In order to minimize the number of urine samples cultured, many laboratories have introduced a preliminary screening of the dipstick type and urines found to be negative are not usually cultured. Support for this action is provided in a hospital based study of 3047 urines tested by N-Labstix[13] where the predictive value of a negative test was 96% and of a positive test was 32%.

The present study analysed the symptoms and history of suspected cases of urinary tract infection, and the result of urine dipstick tests carried out in the practice using N-Labstix. The aim was to increase the precision with which the infection can be identified, thus leading to more rational use of urine culture and effective utilization of antibiotics. In addition the study provided the opportunity to review the occurrence of antibiotic-resistant organisms.

METHOD

The study was conducted at Northfield health centre, Birmingham, during the period November 1984 to June 1985. Three practices took part, involving 10 general practitioner principals and three trainees caring for a total registered population of 18 000 people. Patients presenting with symptoms suggestive of urinary tract infection were recruited. A questionnaire for each patient was completed at the time of recruitment, covering the presenting symptoms (frequency, nocturia, dysuria, haematuria, offensive urine, loin pain, abdominal pain, nausea, incontinence, vaginal discharge); the duration of symptoms; items of history (recent increase in sexual activity, previous urinary tract infection, previous intravenous pyelography, recent catheterization, recent pelvic surgery). Symptoms which were volunteered were recorded separately from those elicited on subsequent questioning. For the symptoms of frequency, nocturia and vaginal discharge a reported increase by the patient was accepted.

The assessment of the doctor of the likelihood of urinary tract infection being present (definite, probable, maybe) and his actions (midstream specimen of urine obtained or not, antibiotic prescribed or not) were recorded. Wherever possible the urine specimen was obtained before beginning antibiotic therapy. Urine was tested at the health centre using N-Labstix for the presence of protein, blood and nitrite and for its alkalinity. The specimens were then forwarded each weekday afternoon to Selly Oak Hospital microbiology laboratory. Specimens from children and pregnant women were all examined by microscopy, culture and sensitivity. All other specimens were first tested using N-Labstix and only examined further if blood, protein or nitrites were present. Further examination included microscopy (cell count in a counting chamber containing mixed but uncentrifuged urine), testing for the presence of bacterial inhibitors and culturing using standard methods and cysteine-lactose electrolyte deficient agar medium. Sensitivity to trimethoprim, ampicillin, sulphonamide, cephalexin, nalidixic acid and nitrofurantoin were assessed routinely when organisms exceeding 100 000 per ml of urine were cultured.

For this study the criteria for infection were either a colony count exceeding 100 000 organisms per ml with a pure urine culture or a count of 10 000–100 000 organisms per ml plus a minimum of 100 leucocytes per mm^3. Urines containing bacterial inhibitors and in which mixed growths were cultured were excluded. The remaining urine samples were classified 'not infected'.

Analysis

After excluding any data set with a major omission (for example, no midstream urine specimen report available, uncertainty about patient identification), remaining data were summarized and entered on a computer file. Data sets in which there were minor omissions (for example, details regarding an item of history or practice dipstick test not available) were processed but excluded in those calculations to which the omission related. Analyses were made in six groups — children aged 0–14 years, pregnant women, and men and women respectively in age groups 15–49 years and 50 plus years.

The occurrence of symptoms in patients with infected urines was compared with that in patients with non-infected urines using a chi-square test with Yates correction. The predictive values (positive and negative) of each symptom, item of history and practice urine dipstick test were calculated. The positive predictive value is the percentage occurrence of these factors in infected urines and the negative predictive value is equal to 100 minus the occurrence in non-infected urines. These values were computed for volunteered symptoms and were compared with the values for volunteered and elicited symptoms combined.

The occurrences in infected and non-infected patients of certain symptoms, items of history and urine dipstick results where there were significant differences between infected and non-infected cases were used to develop a mathematical model in which the probability of a positive diagnosis of urinary tract infection could be calculated (Appendix 1). The score generated in this computation was called the symptoms-history-urinalysis (SHU) score. The percentage of cases predicted using this score was compared with those predicted using a computer aided program[14] to process all items of information obtained in the study. The value of this computer model in assessing the predictive value of symptoms in the management of abdominal pain has been demonstrated.[15] Finally results of the scoring system were compared with the assessment and action of the doctor.

RESULTS

Altogether 521 patients were recruited with suspected urinary tract infection. In 456 patients a midstream specimen of urine was obtained and the laboratory results showed 115 (25%) of the samples were infected, the proportion being greatest among females aged 50 or more years. Of the infected samples 102 contained *Escherichia coli*; other organisms cultured in the 13 other urines were *Proteus* species (3), *Staphylococcus albus* (3), *Klebsiella* species (2), *Streptococcus faecalis* (2), *Staphylococcus aureus* (2) and *Pseudomonas* species (1). Bacterial

Table 1. *Percentage occurrence of symptoms, items of history and urine dipstick test results by patients' group and infection status.*

	Children 0–14 yrs.		Women 15–49 yrs.		Women 50 + yrs.		Men 50 + yrs.		All	
	Infected ($n = 16$)	Not Infected ($n = 59$)	Infected ($n = 46$)	Not Infected ($n = 96$)	Infected ($n = 39$)	Not Infected ($n = 57$)	Infected ($n = 9$)	Not Infected ($n = 32$)	Infected ($n = 115$)	Not Infected ($n = 290$)
Symptoms										
Frequency	62*	28	87*	68	95**	67	58	67	83***	55
Nocturia	31	20	67**	38	87**	49	28	67	64***	36
Dysuria	75**	29	80***	50	78*	49	71*	23	70***	40
Urgency	19	17	39	26	62	42	28	20	43***	24
Haematuria	6	2	18	6	3	7	29	0	14*	5
Offensive urine	12	7	20	15	30	14	28	0	22**	11
Nausea	6	20	4	19	10	20	28	3	9*	19
History										
Symptoms for 9 days or less	93	67	89*	68	59	56	60	32	79**	62
Previous UTI	43	22	52	47	76	54	43	34	57***	40
Previous IVP	14	0	16	6	14	6	40	24	15*	6
Dipstick test										
Protein	56**	10	40**	13	24	27	80*	13	41***	15
Blood	44*	12	74***	29	72	49	100*	25	69***	26
Nitrite	38*	6	46***	0	55***	7	100***	4	54***	3

*$P < 0.05$, **$P < 0.01$, ***$P < 0.001$. Chi-square test (Yates correction) comparing infected and non infected. UTI = urinary tract infection. IVP = intravenous pyelography. n = total number of patients recruited (denominators vary slightly according to the completeness of the information).

inhibitors were found in 26 specimens and mixed growths were cultured in a further 25. Two hundred and ninety specimens were classified as not infected. Of the 115 urine specimens from which organisms were cultured, 67 (58%) contained organisms sensitive to all antibiotics tested. Nalidixic acid was associated with the lowest level of resistance (3%) and ampicillin the highest (28%). The pattern of resistance was broadly similar in all age groups surveyed.

The occurrence of symptoms, items of history and urine dipstick test results in infected and non-infected cases is presented in Table 1 for four of the groups of patients; the numbers of infected urines in pregnant women and men aged 15–49 years were insufficient for inclusion in this table. Items are only shown in the tables where significant differences occurred in one or all of the groups. In the analysis of material relevant to the duration of history a history of nine days or less was found to be the most useful discriminator between the infected and non-infected cases. Table 1 shows that the presence of nitrite in the urine was the single most powerful positive predictor of the presence of urinary tract infection though it was not found in approximately half the infected cases.

The predictive values (positive and negative) of each of these factors among 110 females aged 15–49 years for whom a complete set of data were available were summarized separately for symptoms which were volunteered and for volunteered and elicited symptoms combined (Figure 1). In general, the positive predictive value of a factor improved when elicited symptoms were included but at the cost of a reduction in the negative predictive value. For nocturia and offensive urine there was obvious benefit in including elicited symptoms. Similar results were obtained for children and females aged 50 plus years.

In 263 cases the results of urine dipstick test in the practice could be compared

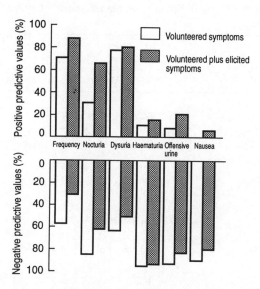

Figure 1. *Predictive values of volunteered symptoms compared with volunteered and elicited symptoms combined in 110 women aged 15–49 years.*

Table 2. *Results of dipstick testing in 263 urine samples: comparison of health centre and laboratory.*

| | Laboratory test | | | |
Practice test	Positive (infected)	Positive (not infected)	Negative	Total
Positive	66	61	20	147
Negative	8	36	72	116
Total	74	97	92	263

with that subsequently undertaken in the hospital laboratory (Table 2). For this purpose any abnormal finding (test positive for protein, blood or nitrite) was accepted as a positive test result. Concordance occurred in 199 (76%) cases. Twenty positive results in the practice were negative on hospital screening and 44 practice negatives were positive on hospital screening; cultures of 36 of the latter samples were not infected.

The occurrence rates of the important items of symptoms, history and urinalysis shown on Table 1 were used to derive the SHU score (Appendix 1). The scoring system is reported for three of the groups of patients (Table 3); the numbers of infected urines in men and in pregnant women were insufficient for this calculation.

Table 3. *The 'SHU' score factors for the diagnosis of urinary tract infection (see Appendix 1).*

	Children 0–14 yrs.		Women 15–49 yrs.		Women 50 + yrs.	
	Present	Absent	Present	Absent	Present	Absent
Symptoms						
Frequency	+2	−2	+1	−3	+1	−3
Nocturia	0	0	+2	−2	+2	−4
Dysuria	+3	−3	+2	−2	+2	−2
Urgency	0	0	+1	−1	+1	−1
Haematuria	0	0	+3	0	0	0
Offensive urine	+2	0	+2	0	+2	−1
Nausea	−2	0	−2	0	−2	0
History						
Symptoms for 9 days or less	+1	−3	+1	−3	0	0
Previous UTI	+2	−1	0	0	+1	−2
Previous IVP	+2	0	+2	0	+2	0
Dipstick						
Protein	+5	−2	+3	−1	0	0
Blood	+4	−1	+3	−3	+1	−2
Nitrite	+4	−1	+11	−2	+5	−2

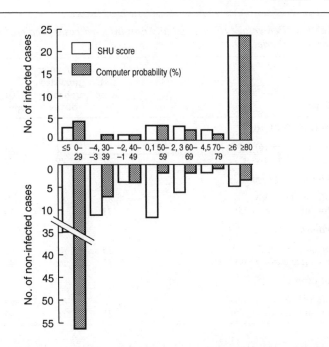

Figure 2. *Numbers of infected and non-infected cases for 110 women aged 15–49 years: comparison of SHU score and computer probability.*

In a separate analysis involving all the items of information collected in the survey, the computer assisted diagnostic model[14] was used to calculate the probability of infection in individual cases. Comparison of the probability assessed in this way with the SHU score for 110 females aged 15–49 years showed both methods produced a similar spectrum of results (Figure 2). An SHU score of zero is equivalent to a probability of infection of 50% and in our opinion represents a suitable level for initiating antibiotic treatment in cases of suspected urinary tract infection. Figure 2 shows that 89% of infected cases and 33% of non-infected cases were included in the group of patients with a positive or zero SHU score; 83% and 11% respectively were included in the group having a computer predicted probability exceeding 50%. For the non-infected group the difference in prediction by the SHU score compared with the computer method (33% as against 11%) was due entirely to the high occurrence rate of several minor symptoms (sweating, loin pain, incontinence, vaginal discharge) in the non-infected group. These differences were not statistically significant and were not included in the SHU score. The differences corresponded to a score factor of −1 for the presence of each of these systems and a score factor of 0 for the absence of each.

A further comparison of the percentage of infected and non-infected cases predicted as infected by the SHU score, the computer and the general practitioner's assessment and action is presented for three patient groups for whom

Table 4. *Percentage of infected and non-infected cases predicted as infected by various criteria.*

Group	SHU score 0+	Computer predicted probability 50%+	GP opinion	GP action (antibiotic prescribed)
Children 0–14 yrs.				
Infected (*n* = 16)	88	81	53	69
Not infected (*n* = 47)	21	19	21	35
Women 15–49 yrs.				
Infected (*n* = 35)	89	83	74	69
Not infected (*n* = 75)	33	11	30	40
Women 50+ yrs.				
Infected (*n* = 27)	85	85	69	81
Not infected (*n* = 47)	15	22	40	35
All cases				
Infected (*n* = 78)	85	83	68	72
Not infected (*n* = 169)	25	16	30	38

complete data are available in Table 4. Computer assisted predictions were marginally better than the SHU score but both were better than the doctor's assessment and action.

The SHU score and computer predicted probability in these three patient groups were also measured in 17 cases in which bacterial inhibitors were found and 19 in which mixed growths were cultured and for whom a sufficient data set was present (Table 5). SHU scores exceeded 0 in five of the cultures with bacterial inhibitors and 11 with mixed growths, and the computer predicted probability exceeded 50% in five and nine cases respectively. Table 5 also includes information about the SHU score and computer predicted probability for patients in which there was non-concordance between urine dipstick testing in the practice and at the laboratory.

The incidence of computer prediction of infection in the mixed growth

Table 5. *Percentage of cases with equivocal culture results for which the SHU score exceeded 0 or the computer predicted probability exceeded 50%.*

	SHU score 0+	Computer predicted probability 50%+
Bacterial inhibitors present (*n* = 17)[a]	29	29
Mixed growths cultured (*n* = 19)[a]	58	47
Non concordant dipstick tests		
Health centre positive, laboratory negative (*n* = 16)[a]	50	38
Health centre negative, laboratory positive (*n* = 36)[a]	18	11

[a]Cases only shown for which a sufficient data set was available.

cases (47%) was higher than that identified in all the non-infected cases (16%, $P < 0.01$) so probability scoring may allow separation of significant mixed growth infection from contamination. In the non-concordant N-Labstix tests, cases found to be positive at the health centre had a higher incidence of computer prediction of infection (38%) than laboratory positive cases (11%), though the difference did not reach statistical significance.

DISCUSSION

Medicine is not an exact science, but is based on probability where actions are determined by identifying favourable risk/benefit ratios. On economic grounds and from obvious commonsense, many laboratories have restricted the number of urine samples cultured though it has to be acknowledged that these are submitted for analysis because of the suspicion of urinary tract infection. Although dipstick testing for protein, blood and nitrite is a useful screen and is entirely appropriate for use in practice it carries an appreciable risk of error when used without any clinical appraisal. Urine culture is the only end-point by which a scientific study of urinary tract infection can be judged. This procedure has limitations, however, as not every urine sample arriving at a laboratory is fresh and uncontaminated, patients sometimes take bacteriostatic agents before consulting the doctor, mixed growths cause confusion and the conventional criterion of 100 000 organisms per ml is itself arbitrary and based on probable rather than absolute criteria.

The key decision for the general practitioner concerns the prescribing of antibiotics, which, if the patient is distressed, ought to be prescribed early rather than late in the illness. Appropriate use of urine culture is desirable but this procedure costs much more than most courses of antibiotics and cannot always be organized easily in rural areas or at inconvenient times of the day or week. Nevertheless there is a tendency by some doctors to overstate this problem; we achieved a midstream urine specimen screen in 87% of suspected cases. The incidence of infection (25%) was lower than that seen in many other reports and this largely reflects the lower threshold for testing patients with suspicious symptoms. In the second national morbidity study 1971–72,[16] the episode rate of urinary tract infection amounted to 38 per 1000 per annum. Applying this estimate we might have expected to recruit approximately 400 patients in the seven months, which compares with the 521 we actually recruited.

A symptoms-history-urinalysis (SHU) score of zero or more identified 89% of infected cases and included only 33% of non-infected cases. This compares well with the computer predicted probability, taking all factors into account, and is substantially better than both the clinical assessment of the doctor and his action, as judged by the decision to prescribe an antibiotic. At the least therefore the SHU score enhances decision-making in cases of suspected urinary tract infection. O'Dowd and colleagues[5] reported a survey in which the management of 46 women with urinary tract infection was compared with that of 40 with the urethral syndrome. They believed these groups could be distinguished if attention were given to psychosomatic factors. These findings differ from those of others,[1,7-9] who could not discriminate on clinical grounds between the two.

When planning the study we were concerned that symptoms volunteered by patients might have a different value than those obtained by direct questioning, a matter of particular importance for computer modelling. Although there were some differences between the predictive values of volunteered symptoms compared with volunteered and elicited symptoms, they were in general small. For the symptoms 'offensive urine' and 'nocturia', there were marked differences which in practical terms favour the inclusion of the elicited symptoms.

The scoring system described translates observed probabilities into a simple summation of single digits to derive an overall probability of diagnosis. It is applied specifically here to urinary tract infection though it can equally be applied to any combination of occurrences in relation to a well-defined diagnosis. The quantification of symptoms, history and urinalysis presented here enhances the quality of the clinical diagnosis and hence the quality of care of patients with urinary problems. Before we are in a position to make logical recommendations for widespread use of the SHU score, this method has to be tested in a prospective survey and this is planned. In the interim we would suggest that an SHU score of zero or more gives a useful guide to diagnosis and would welcome an extension of this approach to, for example, the quantification of symptoms in other areas of medical care.

ACKNOWLEDGEMENTS

We are pleased to acknowledge the cooperation of the general practitioners in Northfield health centre. We are grateful for the contribution made by the laboratory staff at Selly Oak hospital and for the useful comments of Drs I. Craig and K. Kiddy. Finally we are indebted to Dr K. W. Cross in the Department of Social Medicine at Birmingham University for his observations on the scoring method.

APPENDIX 1

Derivation of symptoms-history-urinalysis (SHU) score

The computer based system for combining probabilities of symptoms[14] uses the occurrence rate of each symptom in the infected and non-infected groups. The occurrences of each symptom in the infected group are multiplied together to obtain a product representing the probability that the patient belongs to the infected group (pro-infection product). Similarly, the occurrences of each symptom in the non-infected group are multiplied together giving a product for the probability that the patient belongs to the non-infected group (con-infection product). The actual percentage probability of infection is then obtained by calculating the percentage that the pro-infection product is of the sum of the pro-infection product and con-infection product.

The SHU score system is derived from the computer system to allow simple mental arithmetic calculation of probabilities. SHU-score factors (see Table 3) are derived from the logarithm of the occurrence of the factor in each group, thus allowing addition instead of multiplication of the factor for each symptom. In addition the ratio of occurrence rates in the infected and non-infected groups is used to avoid separate pro and con factors.

$$\text{SHU score} = 2 \times \log_2 \frac{(\text{percentage occurrence in infected cases})}{(\text{percentage occurrence in non-infected cases})}$$

To allow easy mental addition of the scores, logarithms to base 2 are used, and the log value is doubled and rounded to the nearest whole number. This decreases rounding errors and gives scores in the range 0 to ±11. Scores for absence of symptoms are calculated in the same way, using '100 minus percentage occurrence' instead of 'percentage occurrence'.

The final sum is approximately equal to the odds for or against infection for values ±2, ±3, ±4, (for example, a sum of −4 is equivalent to odds of 4 to 1 against infection or a probability of infection of $1/5 \times 100 = 20\%$). For other values, the odds can be calculated as the square root of 2 to the power of the sum (for example for a sum of +6, the odds for infection $= \sqrt{2^6} = 2^3 = 8$ to 1). The final sum is obtained by adding the 'present' factor for any symptom present, and the 'absent' factor for any symptom absent (Table 3). A sum of zero corresponds to a probability of infection of 50%.

REFERENCES

1. Mond, N. C., Percival, A., Williams, J. D., Brumfitt, W. Presentation, diagnosis and treatment of urinary tract infection in general practice. *Lancet* 1965; 1: 574–576.

2. Gallagher, D. J. A., Montgomerie, J. Z., North, J. D. K. Acute infections of the urinary tract and the urethral syndrome in general practice. *Br Med J* 1965; 1: 622–626.

3. Murphy, D. M., Cafferkey, M. T., Faulkiner, F. R., *et al*. Urinary tract infection in female patients—a survey in general practice in the Dublin area. *Ir Med J* 1982; 75: 240–242.

4. Jordan, S., Wilcox, G. M., Wasson, J. H. Urinary tract infection in women visiting rural primary care practices. *J Fam Pract* 1982; 15: 427–428.

5. O'Dowd, T. C., Small, J. E., West, R. R. Clinical judgement in the diagnosis and management of frequency and dysuria in general practice. *Br Med J* 1984; 288: 1347–1349.

6. Brooks, D. A general practitioner's view of the laboratory examination of urine. In: Meers PD (ed). *The bacterial examination of urine; report of a workshop on the needs and methods. Public Health Laboratory Service*. London: HMSO, 1978: 39–44.

7. Steensberg, J., Bartels, E. D., Bay-Nielson, H., *et al*. Epidemiology of urinary tract diseases in general practice. *Br Med J* 1969; 4: 390–394.

8. Brooks, D., Maudar, A. The pathogenesis of the urethral syndrome in women and its diagnosis in general practice. *Lancet* 1972; 2: 893–898.

9. Marsh, F. P. The frequency-dysuria syndrome. In: Blandy, J. (ed). *Urology*. Volume 2. Oxford: Blackwell Scientific Publications, 1976: 734–752.

10. Hamilton-Miller, J. M. I., Brooks, S. J. D., Brumfitt, W., Bakhtiar, M. Screening for bacteriuria: microstix and dip slides. *Postgrad Med J* 1977; 53: 248–250.

11. Zilva, J. F. Is unselected biochemical urine testing cost effective? *Br Med J* 1985; 291: 323–325.

12. Stamm, W. E., Counts, G. W., Running, K. R., *et al*. Diagnosis of coliform infection in acutely dysuric women. *N Engl Med* 1982; 307: 463–467.

13. Lowe, P. A. Chemical screening and prediction of bacteriuria—a new approach. *Med Lab Sci* 1985; 42: 28–33.

14. Horrocks, J. C., McCann, A. P., Staniland, J. R., *et al*. Computer aided diagnosis: description of an adaptable system and operational experience with 2034 cases. *Br Med J* 1972; 2: 5–8.

15. de Dombal, F. I., Leaper, D. J., Staniland, J. R., *et al.* Computer aided diagnosis of acute abdominal pain. *Br Med J* 1972; 2: 9–13.
16. Office of Population Censuses and Surveys, Royal College of General Practitioners, Department of Health and Social Security. *Morbidity statistics from general practice 1971–72. Second national study. Studies on medical and population subjects no. 36.* London: HMSO, 1979.

11 Measuring quality of life

Angela Coulter

WHY MEASURE QUALITY OF LIFE?

Surprisingly few medical therapies have been rigorously evaluated. One pessimistic estimate has suggested that only about 15 per cent of health care interventions are supported by solid scientific evidence (Smith 1991). The 'gold standard' model of scientific evaluation which should be capable of eliminating doubts about the effectiveness of medical interventions is the randomized controlled trial (RCT) (Advisory Group on Health Technology Assessment 1992). Many of the most commonly-used therapies have never been subjected to this type of test, relying instead on observational evidence of risks and benefits.

Observational studies, including non-randomized studies with control or comparison groups, suffer from a major drawback in that one can never be sure that all potential biases have been identified and controlled for. Random allocation of patients to treatment and control groups, if properly conducted, should ensure that biases are evenly distributed between the groups enabling accurate estimation of treatment outcomes. The sample size must be large enough to ensure that the results are not subject to random errors ('the play of chance') and the entry criteria should be broad enough so that the results can be generalized to real-life clinical situations.

Selection of appropriate outcome measures is a key aspect of the design of RCTs. These should be related to both the positive objectives and potential risks of the therapy which is being evaluated. For example if one is planning a trial of a treatment designed to save lives, it is clearly appropriate to measure death rates as a key outcome variable. One should also bear in mind assessment of the quality of the life saved. This becomes increasingly relevant in evaluating the many health care interventions which aim to prevent, cure, or alleviate the effects of diseases and conditions which are not life-threatening. In these cases one wants as a principal outcome to measure the impact of disease and treatment on both health status and health-related 'quality of life'.

Quality of life is a difficult concept to define. In the field of health care evaluation it has come to mean a combination of subjectively-assessed measures of health, including physical function, social function, emotional or mental state, burden of symptoms, and sense of well-being. The key difference between quality-of-life assessment and standard clinical assessment is that it is concerned with the patient's subjective experience of disease and treatment rather than with physiological parameters and physicians' opinions. These two perspectives may be very different and can lead to different valuations of the efficacy of treatment. Clinicians may place more stress on physical outcomes, including physiological indices, survival rates, complications, and recurrence rates whereas for patients the social and psychological impact of treatment may be equally or even more relevant. Since the objective of most therapy in modern health care is to improve or maintain quality of life, it makes sense to monitor outcomes in these terms as well as the more usual clinical outcome measures. This means developing measures of severity of morbidity and of quality of life, both of which have subjective, patient-centred dimensions.

There are now a large number of standardized questionnaires designed to measure health status and health-related quality of life. These measures have been developed in different settings, for different purposes, and using different methods. Some were designed for use with particular patient groups (*disease-specific instruments*) and others are broader measures of health status for use by patients with a wide variety of conditions or problems (*generic instruments*). The stock of standardized instruments includes *health profiles*, which provide separate scores for each dimension of health-related quality of life, and *health indexes*, which give one number for the net effect on quality of life. These latter instruments are still at a rudimentary stage of development but they are favoured by health economists who use them to compare different interventions or policy options by calculating utility-based measures, such as quality-adjusted life years.

Many of these instruments were developed in the USA, where concern about spiralling health care costs has stimulated interest in health care evaluation. The NHS reorganization has prompted similar concerns in Britain and there is now growing interest here in the use of these measures. Pharmaceutical companies have begun to see advantages in incorporating quality-of-life measurement into drug trials. The study featured in this chapter was funded in this way and is an example of the application of a well-established generic instrument, the Sickness Impact Profile (SIP), to assess a treatment for angina. Patients' response to drug treatments is likely to be mediated by aspects of their lifestyle, including social and emotional functioning. If these factors can be measured reliably, it may help in understanding the mechanisms underlying the success or failure of treatments, including the reasons for non-compliance, as well as facilitating the selection of individually-tailored therapies. Another attraction for the drug companies is the possibility that quality of life measures may be more sensitive than the traditional measures of efficacy and may therefore be better able to distinguish differences between similar preparations. The marketing advantage of being able to demonstrate a beneficial effect of a preparation on quality of life has obvious attractions to the industry.

SICKNESS IMPACT PROFILE

The aim of the developers of the SIP was to design a measure of perceived health status that was sensitive enough to detect changes in illness-related behaviour, was applicable across a variety of types and severities of illness, and was culturally unbiased (Bergner *et al.* 1976). It contains 136 statements about health-related dysfunction grouped into twelve areas of activity. These include five independent categories: sleep and rest, eating, work, home management, and recreation and pastimes; three physical categories: ambulation, mobility, body care and movement; and four psychosocial categories: social interaction, alertness, emotional behaviour, and communication (Bergner *et al.* 1981). It can be interviewer or self-administered, taking about 20–30 minutes to complete. Respondents are asked to indicate which statements best describe their current state of health. Each statement or item is allocated a weighted score representing its relative severity. Severity scores were based on the opinions of a panel of judges, including patients and health professionals, who were surveyed during the development of the instrument. The values are summed, divided by the maximum possible score for each category, and multiplied by 100 to give a percentage score. Dimension scores (physical, psychosocial, and total) are calculated in the same way.

A British version of the SIP, entitled the Functional Limitations Profile (FLP), was developed for use in the Lambeth studies of disablement (Patrick and Peach 1989). Although not very different from the SIP, the authors found it necessary to anglicize some of the statements and to re-weight the items using a British panel of judges. Examples of statements in the British version include: 'I do not walk up or down hills' (ambulation), 'I do not bathe myself at all, but am bathed by someone else' (body care and movement), 'I only get about in one building' (mobility), 'I do less of the daily household chores than I would usually do' (household management), 'I go out less often to enjoy myself' (recreation and pastimes), 'I am often irritable with those around me; for example, I snap at people or criticize easily' (social interaction); 'I get sudden frights' (emotion), 'I am confused and start to do more than one thing at a time' (alertness), 'I lie down to rest more often during the day' (sleep and rest), 'I feed myself but only with specially prepared food or special utensils' (eating), 'My speech is understood only by a few people who know me well' (communication), 'I only do light work' (work).

When choosing quality-of-life instruments for use in clinical trials, it is important to select only those that have been tested for *validity* (to ensure that the instrument is adequately measuring the specific dimensions that it is supposed to measure), *reproducibility* or *reliability* (it yields the same results when repeated in subjects whose condition is stable), and *responsive-*

ness or *sensitivity to change* (it is capable of detecting changes over time). The SIP is one of the best measures in these respects. It has been tested extensively, used widely and has been shown to perform well in a number of different settings (Wilkin *et al.* 1992). It correlates well with clinicians' assessments of functioning and it appears to be sensitive to relatively minor changes in health status. Its main disadvantage is that it is relatively long and cumbersome, making it primarily suitable for research settings only.

Some of the other generic instruments for assessing function and quality of life, for example the SF–36 (Ware and Sherbourne 1992; McHorney *et al.* 1993), the Dartmouth Coop charts (Nelson *et al.* 1987), and the Nottingham Health Profile (Hunt *et al.* 1985), are more practical for use in general practice because they are much shorter and therefore take up less of the patient's time. However, the SF–36 and the Dartmouth Coop charts are newer measures which had not been validated on British populations at the time this study was carried out, and the Nottingham Health Profile, widely used in British studies, has been found to be insensitive when used by patients with relatively low levels of morbidity (Kind and Carr-Hill 1987). The SIP was therefore a good choice of instrument for the trial of GTN patches.

QUALITY OF LIFE ON ANGINA THERAPY

The trial described by Fletcher *et al.* (1988) aimed to test the patch method of administering GTN for angina patients. The aim was to see whether this method of continuous administration conferred advantages when used in addition to existing treatments, so the study was organized on a pragmatic basis. Patients continued to use GTN tablets for acute attacks when they needed to and their rate of use of these tablets was used as an indicator of the frequency and severity of attacks during the trial. General practitioners could withdraw their patients from the trial if they felt it necessary. The drawback of the traditional measures of efficacy of angina treatment (attack rates and numbers of GTN tablets taken), as Fletcher and colleagues point out, is that they do not take account of the effect of patients' lifestyle on the development of chest pain. In addition, it was known that GTN carries a risk of headaches. Experience of these side effects could affect use of the tablets and could cause some patients to withdraw from treatment. The investigators wanted to see whether the patches caused the same risk of side effects as tablets and whether there were any other adverse effects of this type of treatment. For these reasons, they decided to supplement the traditional disease-specific measures with two measures of quality of life: the SIP and a health index derived from questions about the impact of symptoms on the patients' quality of life.

STUDY METHODS

Often when we skim medical journals, we skip the methods sections in study reports and jump straight to the more interesting results and conclusions. However, it can be misleading to rely on the authors' interpretation of their findings or to assume that the referees have done a good job in checking that the study design was sound. A careful examination of the methods section can sometimes place a different light on the reported findings. Surprisingly authors often fail to provide the information which the critical reader needs for a full picture of what was done. Altman and Doré (1990) scanned 80 reports of RCTs in four leading medical journals and found that only 34 per cent gave adequate descriptions of the methods of randomization, 34 per cent gave no information on how they decided on the sample size, and the presentation of baseline data was unsatisfactory in 49 per cent.

Unfortunately the present study fails Altman and Doré's test in a number of respects. The authors have provided no details of the methods used to randomize patients into the active and placebo groups nor of the procedure used to ensure that neither patients, doctors nor assessors knew who was in which group (the blinding procedure). This is a serious omission in a trial report, since poorly designed methods of randomization and blinding can lead to biased results. When the authors of the GTN patch trial compared the two groups on baseline characteristics they found that those randomized to the placebo group had worse scores on the SIP than those in the active group. Although it seems likely that this was simply a chance effect, the reader needs information about the method of randomization in order to be sure that this difference did not arise as a result of biased allocation.

Another omission from the description of the methods used concerns the second of the two quality of life instruments. The authors have not provided details of how they adapted Fanshel and Bush's health index for use in the GTN patch trial, making it difficult to evaluate the significance of the results. Since quality of life instruments are probably unfamiliar to the majority of readers of medical journals, it is particularly important that the reasons for their inclusion are clearly stated and the instruments are well described or at least well referenced. Authors of scientific papers face a difficult task because the journals require succinct reports kept within fairly strict word limits. The task will become easier as the instruments are more widely used because it will be possible to refer the reader to documentation of methods of development of instruments and validation of their use published elsewhere. The wide variety of instruments now available means that it should be possible to pick one 'off the shelf' for use in RCTs. Where existing measures do not meet the specific needs of the study, it is essential that new instruments are designed according to the best psychometric principles and tested before use (Streiner and Norman 1989). Quality of life instruments are no different from clinical diagnostic tests in this respect.

STUDY RESULTS

The results of this study nicely illustrate the reasons why double-blind placebo controlled randomized trials are so much more reliable than uncontrolled studies when evaluating treatment efficacy. They also illustrate the value of including quality of life measurement among the outcome measures. All patients showed an improvement in the first four weeks of the study as measured by attack rate and GTN tablet use, but this improvement occurred in the run-in period when patients were not using the patches. During the double-blind period the improvements continued in both the active and the placebo groups, suggesting no difference between the groups and therefore no beneficial effect of the active treatment.

The SIP scores helped to flesh out the picture of the effects of the patch treatment. These showed that there was an adverse effect on the quality of life scores, and in particular the social interaction dimension, for patients on the active treatment. The authors surmise that this was probably due to the increase in the number of headaches experienced by patients when on the active treatment. The usefulness of a *health profile*, such as the SIP, is that it enables an examination of the effects of treatment on the different dimensions of health-related quality of life, thus providing a detailed and clinically relevant estimate of the effects. However, the problem with a multidimensional measure is that it can sometimes show conflicting effects. If, for example, the treatment has a beneficial effect on mobility but an adverse effect on social interaction, the clinician may be faced with a difficult decision on whether or not to recommend its use. A *health index* avoids this problem since it produces only one score for the net overall effect on quality of life, but the gain in terms of simplicity may be achieved at the cost of responsiveness and certainly at the loss of explanatory power. Both types of assessment have a contribution to make; the latter as a crude assessment of the overall effect on quality of life which can be useful when comparing interventions at a policy level; the former as a source of information that can be used to tailor treatments appropriately to the needs of individuals with differing priorities.

We might conclude from this study that the most beneficial effect came from entering patients into a trial and sticking inert patches on them — a relatively cheap and non-invasive form of treatment! If the study had included no control group and no measure of quality of life, the conclusions drawn might have been very different indeed. Without a control group we might have concluded that the improvement in clinical outcomes was due to the treatment. Without the quality of life measures we would not have known about the adverse effects of the active treatment. If the study had not been randomized, we would not have been able to conclude with any certainty that the results were not due to differences between the groups in prognostic indicators.

QUALITY OF LIFE MEASUREMENT IN
GENERAL PRACTICE

When searching for a suitable study to illustrate this chapter on quality of life measurement in evaluative research, I had hoped to find an example of a study which was planned and carried out in general practice, but my search proved fruitless. General practitioners played a key role in this trial since they were responsible for conducting the initial assessment at entry to the study, and for clinical checks at two-weekly intervals. However, the study was organized by epidemiologists on behalf of a pharmaceutical company. It is somewhat surprising that researchers in primary care have been so slow to use the large numbers of tools now available for measuring quality of life. One would have thought that the need for subjective measures would be immediately obvious in primary care, where it is so much more difficult to ignore the relationship between patients' social and domestic situations and their health status. Perhaps the relatively disadvantaged state of general practice research alluded to in the introduction to this book is to blame for the shortage of evaluative studies. David Wilkin and colleagues have published a guide to measuring quality of life in primary care which should help to remedy the situation (Wilkin *et al.* 1992). In this useful book they review 40 different instruments, including measures of functioning, mental health, social support, patient satisfaction, disease-specific measures and multidimensional or generic measures, and describe their applicability to general practice-based studies. Other helpful publications include a BMJ series on quality of life measures in health care (Fitzpatrick *et al.* 1992; Fletcher *et al.* 1992; Spiegelhalter *et al.* 1992) and a guide for users of some of the most popular measures (Jenkinson *et al.* 1993).

The major challenge in primary care research will be to evaluate and compare different forms of disease and problem management as they impact upon patients. This will require a rigorous approach to research design and use of outcome measures which reflect patients' concerns and experience and are clinically relevant. Quality of life instruments have an important role to play alongside physiological measures. It will be necessary for all general practitioners, not just those actively engaged in research, to understand how to interpret studies which have used patient-assessed outcome measures if they are to act on their findings, which after all is the ultimate aim of all evaluative clinical research.

REFERENCES

Advisory Group on Health Technology Assessment (1992). *Assessing the effects of health technologies: principles, practice, proposals*. Department of Health, London.

Altman, D. G., and Doré, C. J. (1990). Randomisation and baseline comparisons in clinical trials. *Lancet*, **335**, 149-53.

Bergner, M., Bobbitt, R. A., Kressel, S., Pollard, W. E., Gilson, B. S., and Morris, J. R. (1976). The Sickness Impact Profile: conceptual formulation and methodology for the development of a health status measure. *International Journal of Health Services*, **6**, 393-415.

Bergner, M., Bobbitt, R. A., Carter, W. B., and Gilson, B. S. (1981). The Sickness Impact Profile: development and final revision of a health status measure. *Medical Care*, **19**, 787-805.

Fitzpatrick, R., Fletcher, A., Gore, S., Jones, D., Spiegelhalter, D., and Cox, D. (1992). Quality of life measures in health care. I: Applications and issues in assessment. *British Medical Journal*, **305**, 1074-7.

Fletcher, A., McLoone, P., and Bulpitt, C. (1988). Quality of life on angina therapy: a randomised controlled trial of transdermal glyceryl trinitrate against placebo. *Lancet*, **ii**, 4-8.

Fletcher, A., Gore, S., Jones, D., Fitzpatrick, R., Spiegelhalter, D., and Cox, D., (1992). Quality of life measures in health care. II: Design, analysis, and interpretation. *British Medical Journal*, **305**, 1145-8.

Hunt, S. M., McEwan, J., and McKenna, S. P. (1985). Measuring health status: a new tool for clinicians and epidemiologists. *Journal of the Royal College of General Practitioners*, **35**, 185-8.

Jenkinson, C., Wright, L., and Coulter, A. (1993). *Quality of life measurement*. Health Services Research Unit, Oxford.

Kind, P., and Carr-Hill, R. (1987). The Nottingham Health Profile: a useful tool for epidemiologists? *Social Science and Medicine*, **25**, 905-10.

McHorney, C., Ware, J., and Raczek, A. (1993). The MOS 36-item Short-Form health survey (SF-36): II. Psychometric and clinical tests of validity in measuring physical and mental health constructs. *Medical Care*, **31**, 247-63.

Nelson, E., Wasson, J., Kirk, J., Keller, A., Clark, D., Dietrich, A., Stewart, A., and Zubkoff, M. (1987). Assessment of function in routine clinical practice: description of the Coop Chart method and preliminary findings. *Journal of Chronic Disease*, **40** Suppl, 55S-63S.

Patrick, D. L., and Peach, H. (1989). *Disablement in the Community*. Oxford University Press, Oxford.

Smith, R. (1991). Where is the wisdom . . .? *British Medical Journal*, **303**, 798-9.

Spiegethalter, D., Gore, S., Fitzpatrick, R., Fletcher, A., Jones, D., and Cox, D. (1992). Quality of life measures in health care. III: Resource allocation. *British Medical Journal*, **305**, 1205-8.

Streiner, D. L., and Norman, G. R. (1989). *Health measurement scales: a practical guide to their development and use*. Oxford University Press, Oxford.

Ware, J. E., and Sherbourne, C. D. (1992). The MOS 36-item Short-Form Health Survey (SF-36) I: Conceptual framework and item selection. *Medical Care*, **30**, 473-83.

Wilkin, D., Hallam, L., and Doggett, M. (1992). *Measures of need and outcome for primary health care*. Oxford University Press, Oxford.

Quality of life on angina therapy: a randomized controlled trial of transdermal glyceryl trinitrate against placebo

Astrid Fletcher, Philip McLoone and Christopher Bulpitt
Epidemiology Research Unit, Royal Postgraduate Medical School, Hammersmith Hospital
The Lancet, 2, 4–7 (1988)

Abstract

In a randomized controlled trial in 427 men with chronic stable angina continuous use of 5 mg transdermal glyceryl trinitrate (GTN) showed no advantage over placebo in terms of efficacy (anginal attack rates and sublingual GTN consumption) or quality of life (as measured with the sickness impact profile and a health index of disability). Patients on the active drug reported headaches more frequently than patients on placebo, and a higher proportion of them withdrew from the trial because of headache. Quality-of-life measurements showed a significant adverse effect of active treatment, principally in the social interaction dimension of the sickness impact profile. A similar effect was observed in placebo patients when crossed to active treatment in a 4-week single-blind period. The results suggest no benefit in the relief of chest pain from 5 mg transdermal GTN when used continuously.

INTRODUCTION

Sublingual glyceryl trinitrate (GTN) is effective in the relief and prevention of angina pectoris, but administration by this route is associated with a high frequency of headaches. Transdermal patches provide a constant dose rate of GTN across the skin, usually over a 24 h period, and several studies have suggested that they are popular with both patients and physicians.[1-3] These studies, however, were both uncontrolled and open. Moreover, traditional measures of efficacy, such as attack rate and GTN consumption, may not reflect the patient's lifestyle – for example, a reduction in the attack rate may be the result of limited physical and social activity. We report here the results of a double-blind randomized controlled trial with transdermal GTN patches in the treatment of angina, using measures of chest pain and quality of life.

PATIENTS AND METHODS

The participants were men aged 30–75 attending their general practitioners with a minimum of 3 months' history of chronic stable angina inadequately controlled on beta-blocker therapy alone. For each patient the dose of beta-blocker was kept constant throughout the study. Sublingual glyceryl trinitrate (GTN) was taken when required for relief of acute anginal attacks, but long-acting nitrates were excluded. All other therapy, including calcium antagonists, was continued. Patients with acute or unstable angina or with a history of myocardial infarction within the previous 3 months were excluded. Patients were randomized at entry to the study to one of two treatment regimens: a 2-week run-in period of observation alone, followed by a double-blind 2-week treatment period with eithers daily placebo patch or a daily transdermal GTN 5 mg patch. All patients then had a single-blind treatment period of 4 weeks on the active transdermal patch. The Rose chest pain questionnaire[4] was completed by the general practitioner at entry to the study. At each assessment (at entry and every 2 weeks) the general practitioner recorded pulse rate and rhythm, sitting systolic and diastolic (phase V) blood pressure, any evidence of cardiac failure, body-weight, concomitant disease and medication, and spontaneously reported symptoms.

Patients were withdrawn from the study if congestive heart failure developed or if they needed other antianginal drugs or were judged by the GP to have adverse effects of therapy.

The effect of treatment was evaluated by means of diary cards and questionnaires. Patients were asked to record each day in the diary card the number of attacks of anginal pain and GTN tablets taken. Diary cards were collected by the general practitioner every 2 weeks. The questionnaire was completed by the patient in the doctor's surgery at entry to the study and at the end of the run-in, double-blind, and single-blind periods. Patients withdrawing from the study at any time were asked to complete the questionnaire. To minimise bias due to information provided by the doctor, all questionnaires were completed before the patient saw the doctor, except those at the beginning of the study, which were completed after the doctor had decided to enter the patient into the trial.

The questionnaire consisted primarily of the sickness impact profile (SIP)[5] with minor adaptations to colloquial English.[6] The SIP consists of 136 statements describing behaviour related to health across a wide variety of dimensions of everyday life. Scores were obtained for each of 12 dimensions and also summarised into physical, psychosocial, and total scores. The psychosocial score has four dimensions: social interaction, alertness behaviour, emotional behaviour, and communication. The physical score consists of ambulation, mobility, and body care and movement. The physical and psychosocial dimensions together with 5 independent dimensions – sleep and rest, eating, work, recreation and pastimes, and home management – produce the final score. High scores imply a poor quality of life. The weights used were those obtained in the English studies (Y. Sittampalam, personal communication) which correspond closely to the original American weights.

Additional questions in the questionnaire included the presence of flushing, headache, and nausea, the number of stairs the patients could climb and distance they could walk, the number of days of incapacity (related either to employment or work around the home), and open-ended questions about the patient's perception of the impact of chest pain or its treatment on lifestyle. A health index was also calculated, giving a score of disability on a continuum from 0 at death to 1 at perfect health. This was modified from previous work[7] and was either derived from responses in the questionnaire or, for patients not completing a questionnaire during the trial, from morbidity and mortality data recorded on the clinical record forms.

Data Processing

Coding of case-record forms, diary cards, and questionnaires and computer entry were done by the statistics unit at Ciba Geigy Pharmaceuticals, Horsham, and checked in a random 10% sample of all data by the Royal Postgraduate Medical School (RPMS) epidemiology research unit. The final data tapes were sent to the RPMS for analysis. Data processing and analysis were done blind to the treatment code.

Statistical Methods

Daily rates of anginal pain and GTN consumption were calculated from 3-point moving averages. Wilcoxon 2-sample tests were used to compare between-treatment scores, between-treatment changes in SIP scores, and the health index.

RESULTS

468 patients from 270 general practices were recruited to the study. 41 patients did not fulfil the entry criteria and were excluded from the analysis. Table I gives the characteristics of patients in the two randomized groups at entry to the study. The SIP and health index scores were significantly different between the two groups. Patients randomized to placebo reported more dysfunction (p < 0.05) and included a higher proportion of patients with a previous myocardial infarction (48% versus 39%).

Angina

98% of patients in each group gave a history of chest pain on walking uphill, and just under 50% in each group gave a history of chest pain when walking on the level. Nearly every patient reported stopping or slowing down when pain occurred, with around 80% in each group reporting less than 10 min as the time for disappearance of pain. Over 75% in each group recorded the sternum or the left anterior chest and left arm as the site of pain. The combination of responses gave angina-positive criterion on the Rose chest pain questionnaire for 67% in the placebo group and 64% in the active treatment group.

Table I *Characteristic of patients at entry to study (mean ± SD or % reporting)*

–	Placebo		Active	
Age	60.4 ± 7.8	(217)	60.5 ± 7.1	(210)
Previous MI	48%	(214)	39%	(208)
Cigarette smoker	17%	(215)	24%	(207)
Beta-blocker with ISA activity	12%	(217)	18%	(210)
Beta-blocker without ISA activity	63%	(217)	69%	(210)
Beta-blocker + calcium antagonist	24%	(217)	23%	(210)
Sickness impact profile				
Physical	8.0 ± 8.2 ⎫		6.8 ± 8.2 ⎫	
Psychosocial*	12.1 ± 12.2 ⎬ (198)		10.0 ± 12.0 ⎬ (191)	
Total*	11.5 ± 9.0 ⎭		9.4 ± 8.9 ⎭	
Health index*	66.3 ± 20.5	(198)	70.1 ± 19.8	(191)

Numbers in parentheses are the numbers of patients for whom the data are available.
ISA = intrinsic sympathomimetic activity.
MI = myocardial infarction.
*p < 0.05.

Table II *Withdrawals during double-blind and single-blind periods*

	Double blind		Single blind	
Reason for withdrawal	Placebo (207)	Active (209)	Placebo →active (198)	Active (195)
Death	1		1	1
Increased angina	2	1	1	4
Headache	1	11	11	3
Skin reaction		1		
Gastric pain	1			
Other and not known	4	1	2	
Total	9	14	15	8
% Withdrawn	4.3	6.7	7.6	4.1

Withdrawals (Table II)

11 patients were withdrawn in the run-in phase (10 in the placebo group and 1 in the active-treatment group). Of these, 3 patients died, 1 had a myocardial infarction, and 1 was admitted to hospital for severe chest pain.

9 of 207 patients on placebo and 14 of 209 patients on active treatment were withdrawn in the double-blind period. 1 patient on placebo died from acute pulmonary oedema. The majority of withdrawals in the active treatment group were due to headache (11, compared with 1 on placebo). The principal reason for withdrawing in the placebo group crossing to active treatment during the single-blind phase was also headache: 15 of 198 patients in this group withdrew (11 for headaches) and 8 out of 195 in the group maintained on active treatment (3 for headaches). 1 death occurred in each group (1 due to a myocardial infarction and 1 during a coronary artery bypass graft operation).

Efficacy

Both the placebo and active-treatment groups showed a fall in the attack rate during the run-in and double-blind phases of the trial, and there was no difference between the two groups in the size of the fall (see figure). A similar result was found for GTN use (figure). The attack rate remained relatively constant in the single-blind period of the trial, although patients crossing over to active treatment tended to have a lower attack rate in the last 2 weeks.

Quality of Life

The changes in SIP and health index scores during the trial (table III) were similar in the two treatment groups during the run-in period, but at the end of the double-blind period patients on placebo had greater improvement in SIP scores, the greatest difference between the two groups being in the social interaction component of the psychosocial score ($p < 0.01$). The change in SIP scores was reduced when placebo patients crossed to active treatment.

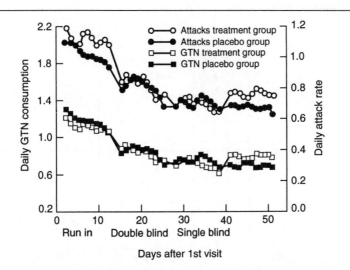

Days after 1st visit

Effects of placebo and active treatment on frequency of anginal attacks and GTN use.

Neither group showed much change in the health index score during the double-blind period, but in the single-blind period patients crossing to active treatment from placebo had a significant deterioration in health index scores compared with those maintained on active treatment.

The average number of stairs the patients could climb and the distance walked improved by a small amount in both groups in all phases of the trial.

To establish whether the change in the social interaction score shown by the placebo group was greater because the initial scores were higher the data were analysed by quartiles of the score at the start of the double-blind period, using the distribution of both groups combined (table IV). The effect of stratification was to make the groups more comparable in each stratum. The upper and lower quartiles showed no difference in the degree of change between the two groups. For patients in the middle two quartiles there was a significant reduction in scores in the placebo group in both quartiles ($p < 0.05$).

Table IV *Mean change in social interaction score in double-blind period stratified by quartiles of score at beginning of double-blind period*

Social interaction at beginning of double-blind period	Change in social interaction		
	Placebo	Active	p
⩾0 to <1.9	−0.6 (2.0)	−0.7 (4.0)	0.9
⩾1.9 to <8.5	0.9 (2.6)	−0.7 (3.8)	0.02
⩾8.5 to <17.9	1.5 (5.3)	−0.3 (4.5)	0.05
⩾17.9	6.3 (13.9)	5.5 (13.2)	0.9

Table III *Mean change in SIP and health index scores (95% confidence interval)*

	Placebo			Active		
Score	Observation, run-in	Placebo, double-blind	Placebo → active single blind	Observation, run-in	Active double blind	Active single blind
SIP scores:						
Physical	0.8 (0.2, 1.4)	0.8 (0.4, 1.2)	0.2 (−0.2, 0.7)	0.9 (0.5, 1.3)	0.3 (−0.2, 0.8)	0.5 (0, 1.0)
Psychosocial	1.6 (0.8, 2.4)	1.8 (0.9, 2.7)*	0.6 (0.0, 1.3)	1.6 (0.8, 2.4)	0.7 (−0.2, 1.6)	1.0 (0.3, 1.7)
Total	1.3 (0.7, 1.9)	1.2 (0.6, 1.8)	0.4 (−0.1, 0.9)	1.2 (0.8, 1.6)	0.6 (0, 1.2)	0.9 (0.4, 1.4)
n	179	183	164	179	186	171
Health index	−0.7 (−2.9, +1.6)	0.1 (−1.8, 1.6)	−2.4 (−4.6, −0.2)*	1.7 (−0.4, 3.8)	−0.3 (−2.0, +1.3)	0.3 (−1.6, +2.3)
n	198	191	192	191	192	192

+, improvement; −, deterioration.
*p < 0.05 for between-drug comparison.

Concomitant disease and symptoms

9% of patients in each group reported one or more symptoms or concomitant diseases during the run-in phase. During the double-blind period these proportions rose to 29% in the actively treated group and 16% in the placebo group. In patients maintained on active treatment the figure fell to 21% during the first 2 weeks of the single-blind period and to 13% during the second 2 weeks. In the placebo group crossing to active treatment the percentage rose to 26% during the first 2 weeks and then fell to 17% during the second 2 weeks. The difference between the treatment groups in these reports was due mainly to headaches. 23% of patients on active treatment spontaneously reported headaches to their general practitioners during the double-blind period, compared with 6% in the placebo group (p < 0.001). During the first and second fortnights of the single-blind phase 11% and 6% on active treatment reported headaches, compared with 20% and 10% of patients crossing to active treatment.

Questionnaire data

Little change was observed in the self-reporting of nausea and flushing by patients in either treatment group. The prevalence of these symptoms was similar in both groups — 14% nausea, 20% flushing. In both groups there was a small increase in headaches during the run-in period. In the double-blind period the reporting of headache increased from 50% to 53% in the active-treatment group, whereas in the placebo group the percentage reporting a headache fell from 51 to 45%.

DISCUSSION

In the large multicentre randomized controlled trial described here 5 mg transdermal GTN was found to have no advantage over placebo in terms of efficacy and quality of life.

Some reports suggest an improvement after 1–4 weeks from transdermal GTN patches applied once daily.[8-14] Other studies have not confirmed a benefit during long-term use,[15-20] possibly because of the development of tolerance.[21] Parker recommended a low-nitrate or nitrate-free period to maintain the antianginal effect.[21] Various studies have shown persistent nitrate effects with continuous therapy.[22-24] Cowen *et al.*[15] demonstrated that antianginal effects were maintained for 1 week when patches were removed each day after 12 h treatment.

In the present trial, both the anginal attack rate and GTN use were reduced when the patch was first applied, whether the patch contained active drug or placebo. Treatment initially seemed to have an adverse effect on quality of life, which was modified as treatment continued.

Inclusion in the trial was itself clearly beneficial in that all measurements of both chest pain and quality of life showed an improvement in the run-in period. The addition of a patch, whether active or placebo, was accompanied by a reduction in the attacks of chest pain.

In the placebo group the change in SIP scores was similar in both the run-in and double-blind periods. However, the improvement in scores of patients with active patches was reduced by at least 50% after 2 weeks of active treatment. This was most apparent for the psychosocial score and social interaction. A similar result was observed when the placebo group crossed to active treatment. This suggests that the perceived benefit of medical care was offset in some way.

In both treatment groups the social interaction scores in patients who reported a headache were double those of patients who did not. Since data on the timing of the headache were not collected, we cannot differentiate between headaches due to additional GTN therapy and those due to transdermal GTN. However, since withdrawals in patients treated with transdermal GTN were mostly associated with headache, and since there was a rise in the prevalence of headaches on active therapy, it is probable that headaches produced by transdermal GTN had a detrimental effect on social function. The twenty statements included in the social interaction dimension cover a variety of aspects such as visiting and family relationships. No single item or cluster of items accounted for the change in score.

A major difficulty, both in the analysis and in the interpretation of the trial results, was the difference in the SIP and health index scores between the randomised groups at entry to the study. The higher scores in the placebo group may have reflected the greater proportion of patients with a history of myocardial infarction—indeed, patients with a history of myocardial infarction had significantly higher scores than those without. However, patients on placebo still had higher scores even when stratified by history of myocardial infarction. The greater changes in SIP and health index scores could have resulted from regression to the mean—i.e., the tendency for initial extreme values to show the greatest change—but no such phenomenon was observed during the run-in period, when the size of the changes was of a similar magnitude in both groups. Moreover, stratification of patients according to the social interaction score produced placebo and treatment groups with similar scores at the start of the double-blind period.

Our results suggest that continuous use of 5 mg transdermal GTN offers no benefit over placebo in the treatment of angina.

ACKNOWLEDGEMENTS

We thank members of the medical and statistics departments at Ciba Geigy, especially Dr Gwen Parr, Mrs Marion Chatfield, and Mr Phil Poole, for their help; Mrs Gail Hazell, of the Royal Postgraduate Medical School, for quality control; Prof Sir David Cox for statistical advice; and the general practitioners who took part in the study for their cooperation. The study was supported by a grant from Ciba Geigy Pharmaceuticals.

REFERENCES

1. Letzel, H., Johnson, L. C. Advances in nitrate therapy for ischaemic heart disease: Results of a multicentre study with Transiderm Nitro. *Z Allg Med* 1983; **59**: 1022–27.
2. Letzel, H., Johnson, L. C. Treatment of angina pectoris with Nitroderm, T. T. S. Results of a multicentre fields study in 37,596 patients. *Med Welt* 1984; **35**: 326–32.

3. Garnier, B., Imhof, P., Spinelli, F., Jost, H. Treatment of patients with angina pectoris in general practice with a new transdermal therapeutic system containing nitroglycerin (Nitroderm, T. T. S.). *Schweiz Rundsch Med Prax* 1982; **71**: 511-16.
4. Rose, G. A., Blackburn, H. Cardiovascular survey methods. Geneva: World Health Organisation, 1986: 56: 1-188.
5. Bergner, M., Bobbitt, R. A., Carter, W. B., Gilson, B. S. The Sickness Impact Profile: Development and final revision of a health status measure. *Med Care* 1981; **19**: 787-805.
6. Patrick, D. Standardization of comparative health status measures: Using scales developed in America in an English speaking community. In: Sudman, S., ed. Health survey research methods. Third Biennial Conference, Hyattsville MD. National Centre for Health Services Research, 1981. PHS Publication no 81-3268: 216-20.
7. Fanshel, S. Bush, J. W. A health status index and its application to health-services outcomes. *Oper Res* 1970; **18**: 1021-26. '
8. Georgopoulos, A. J., Markis, A., Georgiadis, H. Therapeutic efficacy of a new transdermal system containing nitroglycerin in patients with angina pectoris. *Eur J Clin Pharmacol* 1982; **22**: 481-.
9. Thompson, R. G. The clinical use of transdermal delivery devices with nitroglycerin. *Angiology* 1983; **34**: 23-31.
10. Scardi, S., Pivotti, F., Fonda, F., Pandullo, C., Castelli, M., Pollavini, G. Effect of a new transdermal therapeutic system containing nitroglycerin on exercise capacity in patients with angina pectoris. *Am J Cardiol* 1984; **110**: 546-51,
11. Kapoor, A. S., Dazng, N. S., Reynolds, R. D. Sustained effects of transdermal nitroglycerin in patients with angina pectoris. *Clin Ther* 1985; **7**: 674-79.
12. Dickstein, K., Knutsen, H. A double blind multiple crossover trial evaluating a transdermal nitroglycerin system versus placebo. *Eur Heart J* 1985; **6**: 50-56.

Glossary of terms

These are explanatory notes on some of the terms and concepts used in this book, particularly those concerned with statistical treatment of data and with clinical epidemiology. It is not a substitute for the many excellent texts on these subjects, which are listed on page 231.

ALPHA ERROR   The probability that an apparent experimental effect is due to random error. (See *Type I error*).

ATTRIBUTABLE RISK   A measure of the importance of a certain factor in causing a disease, indicating the amount by which the disease burden might be reduced if the causative factor could be eliminated. It is derived from the difference between the *incidence rate* of the disease among exposed individuals and the incidence rate among unexposed individuals, and is expressed as a percentage. It is a guide to individual needs/benefits and may be estimated in *cohort studies* but not in *case-control studies*.

BETA ERROR   The probability that an apparent lack of effect is due to random error. (See *Type II error*).

BIAS   One strict definition is 'any systematic deviation of an observation from the true clinical state', but in most research methods bias refers to any methodological flaw likely to produce systematic deviation from true observation/ measurement, e.g. frequency of symptoms, severity of illness. A **biased sample** is in some way one that is not representative of the population from which it comes. There are three main types of bias: selection bias, information bias, and *confounding*.
   Selection bias operates when factors other than explicit inclusion/ exclusion criteria (e.g. the relationship between investigator and subject) intrude into enrolment. It is found when cases (or controls) are included in (or excluded from) a study on criteria that are related to exposure to the risk factor under investigation. **Berkson bias** is an American term for the important bias(es) introduced in studies of patients drawn only from hospital populations when results are extrapolated to other populations. **Information bias** involves subjects being systematically misclassified according to their disease status, their exposure status, or both. **Interviewer bias** can be of great importance when information is being collected from subjects by face-to-face interview and may be a severe problem when investigators are not '*blind*' to the status (e.g. case or control) of the interviewee. In a questionnaire, inappropriately framed questions, leading questions, and ambiguity may create systematic information bias. **Salience bias** refers to subjects' willingness and ability to recall and provide information on

topics relevant to them, and **recall bias** reflects salience and other factors, such as encouragement and cues from the researcher, influencing subjects' ability to recall information relating, for example, to exposure to risk factors. **The Rosenthal effect** describes selective enquiry and acquisition of data reflecting investigators' expectations of study outcomes. (See also: *confounding*: *non-response bias*).

BLINDNESS A term generally used in clinical trials to describe the lack of awareness of investigators and participants of their inclusion in treatment or control (or traditional/novel) treatment groups in order to eliminate potential bias. In **single blind** trials, the investigator or the subject, but not both, is unaware of whether active or control therapy is being administered. The methodology of **double blind** trials ensures that both the investigators and the participants are unaware of allocation to control or study groups. Blindness is also important in interview studies (see *bias*), when interviewers should generally be blind to the status of the interviewees in the study.

CASE-CONTROL STUDY A retrospective study design in which a sample of people with a particular attribute or disease (*cases*) is compared, with respect to exposure to a certain risk factor, with a sample of people from the same population without the attribute or disease (*controls*), in order to test an aetiological hypothesis. A classic example is the relationship between cigarette smoking and lung cancer. The relative importance of a risk factor is usually expressed as the *odds ratio*, defined as the odds that a case was exposed divided by the odds that a control was exposed, providing an estimate of *relative risk* of disease in exposed compared to unexposed subjects. Case-control studies are relatively cheap and quick, they can investigate a range of current hypotheses and are useful for rare diseases. Beware of biases and get expert help at the design stage if this is your first attempt at this kind of study.

FACTOR	DISEASE Present	Absent	
Exposed	a	b	$a + b$
Not exposed	c	d	$c + d$
	$a + c$	$b + d$	

For a case-control study:

$$\text{Odds ratio} = \frac{\dfrac{a/(a + c)}{c/(a + c)}}{\dfrac{b/(b + d)}{d/(b + d)}} = \frac{a/c}{b/d} = \frac{ad}{bc}$$

CHI-SQUARE TEST   A non-parametric statistical test often used with categorical (cf. ordinal) data in which observed frequencies and expected frequencies are compared, according to the formula:

$$X^2 = \Sigma \frac{(O - E)^2}{E}$$

Where:

O = the observed frequencies
E = the expected frequencies.

The statistical significance of the value of X^2 is determined by reference to statistical tables and the number of degrees of freedom available.

2×2 contingency tables can be used to provide a quick method of analysis of the differences in outcome between two groups, for example

	Drug	Placebo	
Improved	a	b	$a + b$
Not improved	c	d	
	$a + c$	$b + d$	n

$$X^2 = \frac{n(ad - bc)^2}{(a + b)(c + d)(a + c)(b + d)}$$

or, applying Yate's correction

$$X^2 = \frac{n[(ad - bc) - \tfrac{1}{2}n]^2}{(a + b)(c + d)(a + c)(b + d)}.$$

COHORT STUDY   A prospective study design in which a sample of the disease-free population exposed to a risk factor and a sample from the same population not exposed to the factor are followed for a specific period of time and compared for the development of the disease. Doll's study of the development of lung cancer in smoking and non-smoking doctors is a celebrated example of a cohort study. To establish causality, the incidence of the disease under study must be significantly higher in the exposed compared with the non-exposed subjects. Cohort studies are time-consuming, but need to be used to verify hypotheses of causation generated by *cross-sectional* and *case-control* studies.

CONFIDENCE INTERVAL   In a normal distribution, a range of ± 1.96 *standard deviations* on either side of the mean will cover 95 per cent of the area under the curve. For sample sizes greater than about 30, it is possible to calculate 95 per cent confidence intervals to provide information about the range of values within which the true population mean is likely to lie. The 95 per cent confidence intervals are given as the mean value ± 1.96 times the *standard error* (SE); conversely there is a 5 per cent chance of the true population mean lying outside this range ($p<0.05$). There is a 1 per cent chance of it lying outside the range $x \pm 2.58$ SE ($p<0.01$).

CONFOUNDING (see bias) In any study the significance of associations between variables or differences between groups may be undermined by the operation of factors, other than those under investigation, which may themselves explain part or all of the study's observations. Confounding occurs when an estimate of the association between an exposure and disease is mixed up with the real effect of another exposure on the same disease — the two exposures being correlated. In a *case-control study*, failure to control for such an important factor distributed unevenly among cases and controls may distort the real associations between disease and exposure. Elimination of potential confounders is an important part of study design. However, beware the confounder that is not thought of or cannot be adequately measured. It lurks destructively in epidemiological studies.

CONTINGENCY TABLES See Chi-square test.

CORRELATION When one variable changes in a defined way in relation to a second variable, the two are said to be correlated; height in metres is correlated with age in years. Correlation implies association not causality. The correlation coefficient is a numerical expression of the strength and direction (positive or negative) of such an association.

CORRELATION COEFFICIENT A numerical expression of the strength of an association between variables. *Pearson's r* is the appropriate correlation coefficient to use when both variables (x and y) are measured on an interval or ratio scale; r values ≥ 0.7 reflect a high level of correlation and 0.3–0.7 moderate levels. A significance level (p value) may be determined for the range of values of r with different sample $\left(\dfrac{x+y}{2} = n\right)$ sizes.

Spearman's rank correlation coefficient ρ is used when at least one of the variables is not normally distributed. Spearman's rho is given by

$$\rho = 1 - \frac{(6 \times \Sigma D^2)}{n(n^2 - 1)}$$

where n = sample size and D is the sum of the differences in the rank order of the two variables. For sample sizes $n \leq 30$, a special table is needed to determine significance levels; for $n \geq 30$, tables of r are used.

DENOMINATOR In a cross-sectional prevalence study, the prevalence of the condition is given by $\dfrac{n}{D} \times 100\%$ where n = number of cases of the study condition and D = number in the population surveyed; in this equation, D is the denominator and n the *numerator*. In the absence of accurate information about, for example, numbers of registered patients, epidemiological studies may founder on uncertainties about the denominator (*cf.* the UK patient registration system with other health-care systems).

END POINT   An important part of study design is the selection of the main result(s) which will be compared between, say, control and intervention groups, for example survival rates, depression score, consumption of health care resources. Definition of the principal end point(s) at the time the methodology is established enables determination of the sample size required to demonstrate a specified difference between groups in the study, with a specified level of confidence that the difference found did not occur by chance alone. Clarification of end points also avoids problems with post-hoc analysis of a range of outcomes which the study may not have been designed to compare.

FACTORIAL DESIGN   A study design which permits investigation of the separate and combined effects of more than one independent variable on the dependent variable(s); when all possible combinations of the independent variable are used, the study is said to be fully crossed; when not, incompletely crossed. If the effects on a given condition of drug therapy or no drug therapy with or without psychotherapy are studied, the resulting methodology is termed a 2 × 2 design; if a second drug option was added, a 2 × 3 factorial design would result. One advantage of this approach is that increased statistical power can be obtained with relatively small sample sizes, as long as there is not statistical interaction between the variables.

FISHER'S EXACT TEST   When the expected frequency in any cell of a chi-square test is less than five, a special form of the 2 × 2 table should be used to calculate an exact *p* value. First calculate p_1 as follows:-

$$p_1 = \frac{(a + b)\,!\,(c + d)\,!\,(a + c)\,!\,(b + d)\,!}{n!\,a!\,b!\,c!\,d!}.$$

Where ! means factorial i.e. 5! = 5×4×3×2×1. Then reduce the smallest number in any of the cells by 1, to calculate the more extreme probabilities, given again by

$$p_2 = \frac{(a + b)\,!\,(c + d)\,!\,(a + c)\,!\,(b + d)\,!}{n!\,a!\,b!\,c!\,d!}.$$

$P_1 + P_2$ = the probability of finding the observed distribution of data, or a more extreme one.

FOUR BOX (2 × 2) TABLE   A basic tool for the exploration and interpretation of diagnostic tests. *Sensitivity* and *specificity* are stable test characteristics while *predictive values* vary with the *prevalence* of the condition under test.

		Condition under test	
		Present	Absent
Test	Positive	True positive a	False positive b
	Negative	c False negative	d True negative

Sensitivity (true positive rate)

$$= \frac{a}{a + c} \times 100\%.$$

Specificity (true negative rate)

$$= \frac{d}{b + d} \times 100\%.$$

Positive predictive value (predictive value of a positive test)

$$= \frac{a}{a + b} \times 100\%.$$

Negative predictive value (predictive value of a negative test)

$$= \frac{d}{c + d} \times 100\%.$$

Prevalence

$$= \frac{a + c}{a + b + c + d} \times 100\%.$$

FREQUENCY DISTRIBUTION   The way in which scores or values within a given population are distributed. A *normal distribution* produces a unimodal, bell-shaped (Gaussian) curve, in which 95 per cent of the area under the curve lies within the range of the mean ± 1.96 times the *standard deviation*; 99 per cent of the area lies within the range of the mean ± 2.58 times the standard deviation. When data are normally distributed, *parametric* statistical tests may be applied to them; when they are not, *non-parametric* statistics are used.

INCIDENCE RATE   The occurrence of new cases of a disease or condition within a specified population and period of time, for example the incidence of duodenal ulcer in the adult population is approximately 1 per cent per annum (cf. *prevalence*).

INDEPENDENT VARIABLE　In an experimental study, the independent variable refers to the variable factor(s) manipulated by the researcher who determines their effects on the dependent variable(s) i.e. the end point(s) of the study.

INTENTION TO TREAT ANALYSIS　An analysis in which the results in control and treatment groups are analysed with respect to the numbers of patients entering the study (cf. those completing it). By including for analysis all entrants to the study, intention to treat analysis avoids biases due to failure of compliance, and admits side-effects of therapy causing subjects to drop out from therapeutic trials. It provides a pragmatic estimate of the overall benefit of therapy in the population studied.

INTERVAL SCALE　A type of measurement scale in which the values are distinguishable and ordered, the intervals between points on the scale are equal and the zero point is not absolute.

LIKELIHOOD RATIO　= (sensitivity/1 − specificity)

LOGARITHMIC TRANSFORMATION　A method of converting skewed data (i.e. data which are not *normally distributed*) to a normal distribution, by deriving $\log_{10}$ values from logarithmic tables, or by plotting the original data on semilogarithmic graph paper, which has one linear axis and one in $\log_{10}$.

MANN–WHITNEY TEST　Also known as the Mann–Whitney U test and as Wilcoxon's ranking test for unpaired *non-parametric data*. This test determines the statistical significance of the differences between data arising from two different groups of observations. Ranks are assigned to the data as though they all come from one sample; and significance levels for a range of values of rank sums are obtained by reference to statistical tables.

MEAN　The average of a group of values, represented in statistical notation as $\bar{x}$.

MEDIAN　The 'middle' score of a group of values, i.e. the value with equal numbers of other values above and below it. Sometimes used in preference to the mean in describing skewed data because it is less sensitive to the effects of extreme values.

META-ANALYSIS　A method of analysing data from more than one study, with the theoretical advantage of increasing sample sizes. Meta-analysis may detect differences which were not apparent with any confidence in individual studies with small sample sizes. Rigorous criteria for inclusion of data into meta-analyses must be applied to ensure that the data analysed are compatible.

MODE The most frequently occurring value in a group of values.

NOMINAL SCALE Also known as a categorical scale, a nominal scale is one in which the values are distinct categories, for example male or female.

NON-PARAMETRIC DATA Data usually measured on *nominal* or *ordinal scales* for which a *normal frequency* distribution is not observed and for which non-parametric statistical tests should be used for analysis (see: Mann-Whitney test; median values).

NON-RESPONSE BIAS When data from a substantial proportion of a sample are not available for analysis, for example non-responders in a questionnaire survey, information relating only to the responders may not be representative of the population from which the sample is drawn. Non-response bias becomes less important at response rates ≥ 70–80 per cent; below this level, it is particularly important to attempt to define any distinguishing characteristics of non-responders and to consider the possible effects of these on the conclusions of the study.

NORMAL DISTRIBUTION See frequency distribution.

NULL HYPOTHESIS The null hypothesis is an important concept in the design of trials and the understanding of probability. It asserts that all observed variation is due to the variability intrinsic in the material or subjects under investigation. The alternative hypothesis states that a systematic (non-random) influence explains at least part of the observed results such as differences between or relationships between variables. Power is measured by the probability of rejecting the null hypothesis when the alternative is true i.e. that a systematic effect exists.

The investigator evaluates an intervention by posing the null hypothesis and then tests it statistically; if it is rejected at a specific level of confidence, e.g. 5 per cent level, then the alternative hypothesis, that there is a difference between groups or a relationship between variables, is accepted with the corresponding degree of confidence, e.g. 95 per cent.

NUMERATOR The number written above the line in a vulgar fraction. In a population study, for example, the *prevalence* of a condition is given by $n/D \times 100\%$, where n (the numerator) is the number of cases of the condition and D is the *denominator*, i.e. the size of the study population.

ODDS RATIO The odds ratio describes the relative importance of a risk factor investigated in a *case-control study* and is defined as the odds that a case was exposed, divided by odds that a control was exposed (see case-control study). The odds ratio is an accurate estimate of *relative risk* when, in terms of exposure, the disease group is representative of all cases of the disease, the control group is similarly representative of the population of people without the disease, and the *incidence* rate of the disease in the general population is very low.

ONE TAILED TEST   When a difference between groups of data is expected to occur only in a particular direction a one-tailed statistical test is used. In practice this rarely applies. (A two-tailed test is used if the statistical difference between the groups is tested without reference to the direction of the difference).

ORDINAL SCALE   A series of values which are distinguishable and ordered but the intervals between them are not necessarily equidistant, the values of an ordinal scale are often ranks, for example 2nd, 4th.

PARAMETRIC DATA   Data for which a normal *frequency distribution* is assumed, which are measured on *interval* or *ratio* scales and for which parametric statistical tests are appropriate.

PEARSON'S CORRELATION COEFFICIENT (*r*)   A measure of the extent to which paired scores, measured on interval or ratio scales and assumed to be normally distributed, are correlated. The closer the relationship between the variables approximates to a straight line, the more *r* approaches + 1. The significance (*p* value) of values of *r* for different sample sizes (*n*) can be obtained by reference to statistical tables (see *correlation coefficient*).

POWER   The probability of detecting that the presence of a specified level of effect (e.g. a difference in outcome between placebo and treatment groups) is not due to chance alone. Power is defined mathematically as 1 − *Beta error*. Power calculations involve the determination of sample sizes required to detect an effect at the desired level of significance (e.g. $p < 0.01$), i.e. with specified confidence that the effect was not found by chance alone. (See also *Type II error*).

PREDICTIVE VALUE   (Positive and negative) See *four box table*.

PREVALENCE   The proportion of cases within a specified population at a given time; point prevalence signifies a current, 'snapshot' prevalence, whereas **period prevalence** denotes the proportion of cases in the population within a specified observation period (cf. *incidence*, see also *four box table*).

QUALITATIVE METHODS   A research approach in which non-numerical, e.g. narrative and interpretive analysis is emphasized in the study of social phenomena (cf physical science) and the contexts in which they appear. The approach seeks to describe culture, beliefs, and attitudes, and to illuminate and understand behaviour rather than to predict it. It can also lead to hypothesis generation for epidemiological testing.

QUANTITATIVE METHOD   A research approach which concentrates on the collection of numerical data which are subject to statistical analysis to test hypotheses generated by the researcher, for example in the study of aetiology and the public health impact of disease.

QUOTA SAMPLING A sampling procedure in which quotas are set for various categories, so that a sample of 60 children might have quotas of 20 aged under 5, 20 from age 6–10 and 20 aged over 11.

RANDOM SAMPLING A sampling procedure in which cases are drawn randomly from the population, of which each member has an equal chance of selection.

RANDOMIZED CONTROLLED TRIAL (RCT) The most powerful experimental design for obtaining evidence of both causation and public health impact. Subjects are randomly allocated to exposure to the intervention of interest or to a control or comparison group. The differences in outcome (*end points*) between the two groups are compared and tested for significance or estimates of difference are defined by confidence intervals. (See also *blindness, cohort study, case-control study*).

RATIO SCALE A measurement scale in which the values are distinct, ordered, and equidistant, and where the zero point represents an absence of the quantity measured. Metres and kilograms are examples of ratio scales.

RELATIVE RISK The ratio of the incidence of disease in exposed persons to the incidence in non-exposed persons (see also *cohort study, case-control study*).

RESPONSE RATE The proportion of a sample of subjects responding to an invitation to participate in a study (e.g. to complete a questionnaire or to attend for a screening test), expressed as a percentage of the sample selected.

SENSITIVITY The true positive rate of a diagnostic test (see *four box table*).

SPECIFICITY The true negative rate of a diagnostic test (see *four box table*).

STANDARD DEVIATION (SD) A measure of the variability of a group of normally distributed scores, defined as the square root of the variance, expressed as mean ± SD ($\bar{x}$ ± SD).

STANDARD ERROR OF THE MEAN (SE) A measure of the deviation of scores from the estimated population (cf. sample) mean, defined as the variance divided by the square root of the sample size.

STRATIFICATION A sampling method in which individuals for study are selected from within subgroups of a population rather than sampling from the entire population (*quota sampling*). Stratification is carried out in order to ensure representativeness or exclude bias.

t-TEST   A statistical test for determining the significance of differences between sample means of normally distributed data; may be *one-tailed* or *two-tailed*.

TWO BY TWO TABLE   See *four box table.*

TWO-TAILED TEST   A statistical test for the difference between two groups without reference to the expected direction of that difference; usually employed in pragmatic studies.

TYPE I ERROR (*ALPHA ERROR*)   The error, in the analysis of data, of stating that a difference/effect is present when in fact it is not; conventionally the numerical value for alpha is often set at 0.05, i.e. on 5 per cent of occasions the size of difference found could have occurred entirely by chance.

TYPE II ERROR (*BETA ERROR*)   The error, in the analysis of data, of concluding that an effect/difference is not present when in fact it is; a particular problem when sample sizes are small. Conventionally, beta is often set at 20 per cent, signifying a 20 per cent chance of missing a true effect. Conversely, such a study is said to have 80 per cent *power* for detecting an effect of the magnitude specified.

U TEST   See *Mann-Whitney test.*

VARIANCE   A measure of the dispension or variability around the mean of a group of scores, represented by the symbol s^2 and calculated as

$$s^2 = \frac{\Sigma(x - \bar{x})^2}{(n - 1)}$$

where $(x - \bar{x})$ is the difference between each individual score and the mean and n is the sample size.

YATE'S CORRECTION   When the chi-square test is used to test differences between groups, a correlation for continuity (Yate's correction) is usually applied for sample sizes of $n < 100$.

Suggestions for further reading

Armstrong, D., Calnan, M., and Grace, J. (1990). *Research methods for general practitioners* Oxford University Press. ISBN 0-19-261822-9.

Hennekens, C. H., and Buring, J. E. (1987). *Epidemiology in medicine.* Little, Brown and Co., Toronto.

Howie, J. G. R. (1989). *Research in 'general practice* (2nd edn). Croom Helm, London. ISBN 0-412-33730-4.

Norton, P. G., Stewart, M., Tudiver, F., Bass, M. J., and Dunn, E. V. (1991). *Primary care research: traditional and innovative approaches.* Sage Publications, Newbury Park. ISBN 0-8039-3871-3.

Pocock, S. (1983). *Clinical trials: a practical approach.* John Wiley and Sons Chichester.

Polgar, S. and Thomas, S. A. (1991). *Introduction to research in the health sciences.* Churchill Livingstone, Melbourne. ISBN 0-443-04363-9.

Sackett, D. L., Haynes, R. B., Guyatt, G. H., and Tugwell, P. (1991). *Clinical epidemiology: a basic science for clinical medicine,* (2nd edn). Little, Brown and Co., Boston. ISBN 0-316-76599-6.

Selvin, S. (1991). *Statistical analysis of epidemiologic data.* Oxford University Press, New York. ISBN 0-19-506766-5.

Siegel, S. and Castellan, N. J. (1988). *Nonparametric statistics for the behavioural sciences* (2nd edn). McGraw-Hill, New York. ISBN 0-07-057357-3.

Stewart, M., Tudiver, F., Bass, M. J., Dunn, E. V., and Norton, P. G. (1992). *Tools for primary care research.* Sage Publications, Newbury Park. ISBN 0-8039-4404-7.

Index

Guidelines for critical reading (From Sackett *et al.* 1991 with permission).

Is the article a report of an original study or a critical review that is directly relevant to your own clinical practice? **NO** ──────→ GO ON TO NEXT ARTICLE

YES ↓

Purpose of study?

Therapy	Diagnosis	Screening	Prognosis	Causation	Quality of Care	Economics Analysis	Review
Was the assignment of patients to treatments really randomized?	Was the test compared blindly with a gold standard?	Was the study a randomized trial? If YES, see Therapy If NO:	Was an inception cohort assembled?	Was the type of study strong? (RCT > cohort > case control > survey)	Did the study focus on what clinicians actually do?	Did the economic question include alternatives to be compared and viewpoint?	Were the questions clearly stated?
Were clinically important outcomes assessed objectively?	Was there an adequate spectrum of disease among patients tested?	Are there efficacious treatments for the disorder?	Were baseline features measured reproducibly?	Was the assessment of exposure and outcome free of bias? (e.g. blinded assessors)	Have the clinical acts under study been shown to do more good than harm? If not, did the study compare process to outcome?	Were the alternative programs adequately described?	Were the criteria for selecting articles for review explicit?

Was the treatment feasible to use in your practice?

Was the referral pattern described?

Does the current burden of suffering warrant screening?

Were the outcome criteria clinically important and reproducibly measured?

Was the association both significant and clinically important? If not, was power considered?

Were the clinical processes or acts measured in a clinically sensible and valid way?

Have the program's effectiveness been described?

Was the validity of the primary studies assessed?

Was there at least 80% follow-up of subjects?

Was the description of the test clear enough to reproduce it?

Does the screening test have high sensitivity and specificity?

Was follow-up at least 80%

Was the association consistent across studies?

Were both clinical and statistical significance considered?

Were all relevant costs and effects identified?

Was the assessment of primary studies reproductible?

Were both statistical and clinical significance considered?

Was the test reproducible (observer variation)?

Can the health system cope with the screening program?

Was there adjustment for extraneous prognostic factors?

Was the 'cause' shown to precede the 'effect'?

Were the measurements credible?

Was variability in the results of studies analysed?

If the study was negative, was power assessed?

Was the contribution of the test to the overall diagnosis assessed?

Will positive screenees comply with further assessment and intervention?

Was there a dose-response relationship?

Was a sensitivity analysis performed to assess the effect of assumptions?

Were the findings of the primary studies combined appropriately?